Biomedicine

Biomedicine

Sidney Ashworth

MURPHY & MOORE
www.murphy-moorepublishing.com

Published by Murphy & Moore Publishing,
1 Rockefeller Plaza,
New York City, NY 10020, USA
www.murphy-moorepublishing.com

Biomedicine
Sidney Ashworth

International Standard Book Number: 978-1-63987-077-6 (Hardback)

Cataloging-in-Publication Data

Biomedicine / Sidney Ashworth.
p. cm.
Includes bibliographical references and index.
ISBN 978-1-63987-077-6
1. Medicine. 2. Biomedical engineering. 3. Medical sciences.
4. Biology. I. Ashworth, Sidney.
R852 .B56 2022
610.28--dc23

Table of Contents

Preface

This book is a culmination of my many years of practice in this field. I attribute the success of this book to my support group. I would like to thank my parents who have showered me with unconditional love and support and my peers and professors for their constant guidance.

The branch of medical science that uses physiological and biological principles to clinical practice is called biomedicine. It deals with a wide variety of technological and scientific approaches such as in vitro diagnosis and fertilization. The field focuses on standardized evidence-based treatment that has been validated through biological interventions. It includes the study of physiological processes by using methods drawn from physiology and biology. The discipline involves other areas such as biological engineering, cell biology, gene therapy, biotechnology, systems biology, embryology, molecular biology, nanobiotechnology, biochemistry, laboratory medical biology and other life sciences that are applied to medicine. While understanding the long-term perspectives of the topics, the book makes an effort in highlighting their impact as a modern tool for the growth of the discipline. The various sub-fields of this discipline along with technological progress that have future implications are glanced at in this book. Constant effort has been made to make the understanding of the difficult concepts of biomedicine as easy and informative as possible, for the readers.

The details of chapters are provided below for a progressive learning:

Chapter - What is Biomedicine?

The branch of medical science that uses physiological and biological principles in clinical medical practice is termed as biomedicine. It includes the use of stable magnetic isotopes and optical fiber gratings. This chapter has been carefully written to provide an easy understanding of biomedicine.

Chapter - Biomedical Engineering

Biomedical engineering is an interdisciplinary field of engineering and medicine that applies engineering design and principles to medicine and biology for diagnostic and therapeutic purposes. Biomimetics, neural engineering, bioinformatics, biomechanics, etc. fall under its domain. This is an introductory chapter which will briefly introduce about biomedical engineering.

Chapter - Medical Biology

The branch of biology that deals with the practical applications of medicine, health care and laboratory diagnostics is called medical biology. It includes other disciplines such as molecular biology, nanobiotechnology, cytogenetics, microbiology, toxicology, etc. This chapter discusses the subject of medical biology in detail.

Chapter - Medical Genetics

Medical genetics is the branch of medicine that is concerned with the management and diagnosis of hereditary disorders. Genetic engineering, gene therapy, genetic testing and counseling, etc. are studied under its domain. This chapter closely examines medical genetics to provide an extensive understanding of the subject.

Chapter - Molecular Medicine

The field of medicine that uses physical, chemical, bioinformatics and medical techniques for describing molecular structures, identifying molecular and genetic errors of disease, and developing molecular mechanisms. All the aspects related to molecular medicine have been carefully analyzed in this chapter.

Sidney Ashworth

1

What is Biomedicine?

The branch of medical science that uses physiological and biological principles in clinical medical practice is termed as biomedicine. It includes the use of stable magnetic isotopes and optical fiber gratings. This chapter has been carefully written to provide an easy understanding of biomedicine.

Modern western scientific medicine is often called "biomedical" because it explains health in terms of biology. It attaches importance to learning about body structure (anatomy) and systems (physiology), in particular to understanding mechanisms like the heart, arteries, nerves, brain and so on.

Health is seen as a state where all the parts of the body function normally, like a new repaired car. If bits go wrong–if the body is struck down by a virus, internal changes damage it, or parts wear out–it goes in for repair by specialists. This view offers a particular and distinctive way of 'seeing' and understanding bodies and health, a set of guidelines about relating to them and dealing with them–to look after them as systems which need care and proper maintenance by their owners.

So a biomedical account is one which gives a physical or biological explanation for health, and offers physical/biological methods for 'repairing' bodies when they are not working correctly. Certain tests establish what is wrong. Then antibiotics or other medicines act as a cure, or surgery can repair or replace body parts. The biomedical model is an integral part of western cultures and the way health and healthcare are perceived. It is, in many instances, an efficient and effective model of healthcare (consider broken legs, tumours, tuberculosis, slipped discs and a host of other illnesses or physical problems).

People's views of their health, how it can be maintained and the appropriate action to combat illness when it is identified are, of course, individual. However, individual understandings emerge partly from wider historical and cultural patterns and trends. It is therefore worth reflecting on how the biomedical way of conceptualising health has come to dominate in the West.

Biomedicine, Public Health and Germ Theory

With the rise of biomedicine came greater knowledge about the causes of infectious diseases. With greater urbanisation during the Industrial Revolution came widespread disease, which led to the public health movement. The key tenet of public health theories, of whatever type, was (and is) that health and disease arise from the relationship of individuals and populations with their natural or manufactured environment, and that the promotion of health and wellbeing requires intervention to modify or transform that environment.

In the late nineteenth century, however, the significance of public health theory was challenged by the germ theory of disease which suggested that disease was caused by bacteria, and that a specific micro-organism would be present in every case of a particular disease. Most medical research was by now being done within a dominant scientific paradigm. Germ theory was derived from nineteenth century microscopic research by Louis Pasteur and Robert Koch, among others. Pasteur demonstrated the existence of micro-organisms, which could be killed by sterilisation and the exclusion of 'impure air', or made to multiply rapidly in suitable media. Koch succeeded in isolating the anthrax bacillus, growing it in culture, and proving by experiment that its spores could reproduce rapidly even after a long delay, when conditions became favourable. This provided the other principles: that the micro-organism should be able to be grown outside its host, and should still be capable of producing the disease again in the host body.

Germ theory resulted in an enhanced status for laboratory science, and for hospital medicine and surgery. Laboratory scientists such as Louis Pasteur, Marie Curie, James Lister and Robert Koch could isolate and classify bacteria; medical people could use this knowledge to diagnose more accurately and operate more safely. Germ theory focused attention on the specific aetiology and the immediate cause of disease. It also focused attention on the individual and on interpersonal contact, rather than on the significance of the wider social environment in promoting health. Within the public health field, greater attention was given to personal hygiene and to educating working-class women in mothering skills. Germ theory reinforced the official attack on personal filth—spitting, lice, scabies, dirty bodies—but provided no commentary on the social conditions which made such problems widespread in poorer districts.

Germ theory and laboratory science remain important today as doctors rely more and more on laboratory testing to help them diagnose and monitor specific conditions or diseases. However, specific aetiology (that is, the same micro-organism causing the same illness) became increasingly open to question by the mid-twentieth century as scientific research revealed the significance of co-factors, host state, as well as invading organism. In the 1970s Engel called for a 'new medical model' which emphasised multi-causality. Research into AIDS, for example, has produced theories which emphasise multiple causes.

Biomedicine has been a dynamic model and played a hugely important part in people's understandings of health and illness and their compliance with the system that supports biomedicine. Proponents of biomedicine suggest that it stands with many fantastic accomplishments: the reduction of high mortality rates such as those found in childhood in the earlier part of the twentieth century, and the eradication of some of the major killers, such as smallpox, polio etc. However, an opposing argument has been put forward that mortality has been dramatically reduced during the past century because of improved housing and working conditions and particularly improved nutrition, rather than because of any particular preventive or biomedical measure, such as immunisation or the widespread use of antibiotics. Whereas biomedical discourse tends to focus our attention on medical advances and interventions at an individual level.

Biomedicine approaches the relationship between religion and healing from a number of perspectives. The first involves the religiously informed bioethical dilemmas that arise in relation to medical decision making, sometimes in relation to religiously based prescriptions or prohibitions of actions or substances. The most routinely cited are the rejection of blood transfusions by Jehovah's Witnesses and of biomedical interventions by practitioners of Christian Science. In such cases, religion is represented primarily as an 'unscientific' obstacle to biomedical therapies, and not as a source of healing in its own right.

Technological innovation often drives bioethical debate, rather than ethics informing the development of technology. Rather than reflecting the pluralism within religious traditions, bioethics tends to present traditions as having generic static, and normative ideals for correct action, rooted in the interpretation of authoritative sources. This focus bypasses accounts of the customized, lived ways in which individuals and communities actually integrate authoritative rulings (to the extent they are aware of them) with other factors, when making ethical decisions in particular social and personal contexts.

Another approach focuses primarily on quantifiable health outcomes and on the less tangible factor of religious coping in the face of illness or other forms of suffering. Although individual biomedical physicians may craft a host of personalized ways to integrate their religious lives with their medical practice, biomedicine as a discipline has resolutely distanced itself from religion, which has been framed as antithetical to science. Yet not all researchers have accepted the mutual exclusiveness of the two domains. Some have applied medical research models to the impact of religiosity on health, in the hope that by employing standards of research and evidence most familiar to, and accepted within, biomedical culture, their colleagues may become less reluctant to acknowledge and build on links between the domains.

This undertaking has posed the challenge of formulating an operationalized definition of religiosity that can be articulated through scales. Many such scales have focused on religious affiliationor membership, religious participation or attendance at formal services, private religious practices (e.g., prayer, meditation, reading sacred literature), and religious coping in response to difficult circumstances. Critics argue that few of these scales truly represent the religious intent of practitioners or religious pluralism, reflecting instead the often Christian cultural paradigms of their designers. Such scales may be well suited for the study of some Christian communities, but fall short when investigating most other religious traditions. Alternate approaches might, for example, focus on ritual, the observance of prohibitions and prescriptions, and other dimensions of practice.

Biomedical approaches have also examined ways in which religiosity may serve as a protective factor against negative health outcomes. Examples include ways in which religious communities may help individuals internalize healthy lifestyle choices and provide external sanctions for violating norms. Religious coping strategies and patterns of mutual economic and social support likewise receive attention. Some links between various indicators and specific coping strategies have been suggested, but variables of social location, types of health problem, as well as the diversity and changeability over time of religious orientation limit their predictive value and generalizability. Critics often challenge the design of such studies and argue that few clinicians have adequate training to serve in the capacity of spiritual advisor or counselor, particularly in the face of growing religious pluralism.

The rise of biomedicine not only in the West, but also around the world, has led to its routinely being referred to simply as medicine. Yet this apparent triumph of a single system belies the historical and contemporary pluralism of healthworlds and healing practices integrated with biomedicine in these same contexts. Few, if any, communities or cultures are homogeneous, whether in their religious traditions and practices, or in their approaches to health, healing, and related interventions.

In health decision making, multiple rationalities come into play, whether the process be conceptualized as following hierarchies of resort or patterns of resort. Whether such practices are characterized as folk, ethnomedical, traditional or, most recently, as alternative, complementary, or integrative, they reflect the underlying reality that many people routinely use multiple healing practices. Indeed, those who do not are often the exception. Some of these practices emerge from religious worldviews and traditions, whether overtly or less directly, although they are rarely included in the definitions of religiosity. Studies of complementary/alternative medicine (CAM) in the United States wrestle with whether or not to include practices such as prayer. However, by including prayer – whether personal or intercessory – the numbers of CAM users across demographic categories rise exponentially.

Increased data revealing significant disparities in disease burden, access, and outcome between patients, based on gender, class, race, and ethnicity, have led to initiatives to train clinicians to become culturally competent, capable of negotiating across differences in patient understandings of illness, health, and intervention. Yet the requisite skills and frameworks differ from other professional competencies, particularly insofar as they often entail addressing otherwise unexamined ignorance, indifference, and biases in relation to cultural and spiritual aspects of care. Although urban health-care institutions in the West have begun to recognize obstacles to appropriate care for religious minority patients who may face discrimination related to religious difference, the ongoing challenge remains one of not reducing culture or religion to lists of stereotyped traits. Rather, health educators, providers, and their clients more often live in contexts of hyperdiversity; where multiple ethnic, cultural and religious identities intersect within individuals, families and other social groups.

STABLE MAGNETIC ISOTOPES IN BIOMEDICINE

Diverse organisms possess the ability to perceive Earth's magnetic field, the strength of which is about 0.05 mT. There exist magnetotactic bacteria the ability of which to use geomagnetic fields for direction sensing is accomplished owing to the so-called magnetosomes, the specific nanometer-sized magnetite particles organized into chains within the cell.

However, apart from external magnetic fields, another variety of natural magnetism is around, namely, magnetic fields of atomic nuclei of magnetic isotopes. Some of them produce intramolecular magnetic fields which are 10-100 times greater than terrestrial. This raises the question of whether living cells can perceive the difference between magnetic and non-magnetic isotopes of the same chemical element. There is also a practical issue of whether the cell can take advantage of the magnetic isotopes.

Ab Initio: Magnetic-isotope Catalysis in Chemistry and Biochemistry

Apart from the energy control (the law of conservation of energy), any chemical reaction as electron-nuclear rearrangement of reactants into products is controlled by angular momentum, spin, of reactants. Namely, the total spin of reaction products must be identical to the total spin of reactants. This law of spin conservation immediately follows from quantum mechanics, from the fundamental and universal Pauli principle: no two electrons may occupy the same quantum state simultaneously.

Figure illustrates how the law of conservation of spin gives control over reactivity of free radicals, Rï. For example, a pair of free radicals, each with electron spin S=1/2, may form a chemical bond and the resultant diamagnetic molecule, the total electron spin of which S=0. However, from the law of conservation of spin, it follows that the chemical bond between these two radicals may happen only if the spin state of the pair at collision is singlet, i.e., spins of two electrons are subtracted to give the net S=0 (spin multiplicity, 2S+1=1). If the spin state of the radical pair is triplet, i.e., if the electron spins are added up to give the net S=1 (spin multiplicity, 2S+1=3), then the radicals cannot react immediately.

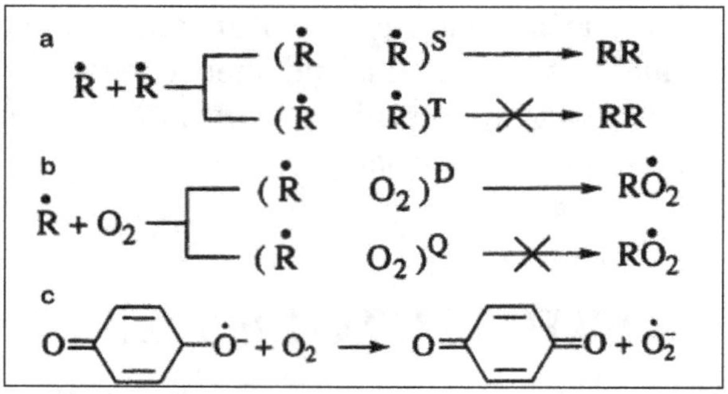

Spin control in the chemical reactions of free-radical pairs.

In gas or liquid phase, with the time allotted for any collisions of radicals of order of 1 ns or less, neither spin-spin relaxation (in order of 10 ns) nor spin-lattice relaxation (in order of 100 ns) has time to fit the spin orientation. As a result, only one-quarter of encounters, with the radical pair in the singlet state, gives the recombination product while three-quarters of the initial radical pairs are inhibited from the reaction. Another example is presented on figure. Namely, it is the reaction of free radical Rï with oxygen, molecules of which are normally in the triplet spin state. The total spin, S, of this reagent pair can be 1/2 when the individual spins are subtracted (2S+1=1) or 3/2 when the individual spins are added up (2S+1=4). Meanwhile, for the product of the reaction RO_2 ï, peroxyl radical, S=1/2. Hence, from six possible spin states of the reactants only two states do not require a change in the total electron spin of the reactants and, therefore, are permitted for formation of RO_2 ï; other four states are forbidden for the reaction. A well-known reaction of mitochondrial ubisemiquinone with oxygen, in which O_2 ï– is produced, exhibits a similar case. This reaction is permitted only from the doublet state of the reactants. Four quartet "channels" are forbidden by the law of the spin conservation.

To lift the ban on reactions forced by the law of spin conservation, spins of the reactants must be changed. Inasmuch as spin-orbital coupling is negligibly small in organic free radicals, magnetic fields are the only means which are able to change the spin states and, thereby, switch the reaction over spin-forbidden and spin-allowed channels. The probability of chemical reaction is a function of the parameters of magnetic interactions.

$$P = f(H; or; H_1; J; a; I; m_I; \mu_n)$$

In this equation H is external magnetic field (Zeeman interaction), ω and H1 are frequency and amplitude of microwave magnetic fields. Correspondingly, acceleration of the freeradical reaction can be achieved through changes in the total electron spin of reactants by interaction with an applied external magnetic field. The parameter J is energy of the exchange interaction. Correspondingly, the reactions of organic free radicals or ion-radicals can be catalyzed via interaction of partners of the radical pair with a foreign, third spin carrier, like nitroxide radical. It is called "electron spin catalysis".

The above mentioned equation also contains parameters of hyperfine coupling a, nuclear spin I, nuclear spin projection mI, and nuclear magnetic moment μn, i.e., the parameters of interactions of electron spins with magnetic nuclei which are known as the cause of the hyperfine splitting in EPR spectra of free radicals. Correspondingly, acceleration of the freeradical reactions can be achieved through changes in the total electron spin of reactants by interaction with magnetic fields of magnetic nuclei. This is known as "magnetic-isotope effect" (MIE): the reaction shows different reaction rates and different yields of products according to whether the reagents contain magnetic or nonmagnetic isotopes. While classical isotope mass effect selects isotopic nuclei in accordance with their masses, MIE selects isotopes by spin and magnetic moment. In action, MIE is a purely kinetic phenomenon and manifests itself as the dependence of the reaction rate on the nuclear spins of the reactants. Within recent years, MIE in chemistry has been discovered for a number of magnetic isotopes, among them H–D, ^{13}C, ^{17}O, ^{29}Si, ^{33}S, ^{73}Ge, $^{117,119}Sn$, $^{199,201}Hg$, and ^{235}U. By analogy with "electron spin catalysis", the enhancement of the reaction rate by the nuclear spins of the reactants can be denoted as the "nuclear spin catalysis".

In biochemistry, MIE has been recently discovered for magnetic isotope of magnesium, ^{25}Mg, by A.L. Buchachenko and his group. It is generally known that energetic demands of every operation in living systems are met by molecules of ATP, be it eukaryotic cells of animals and plants or prokaryotic cells of bacteria. In aerobic organisms, most of ATP is produced in the so-called "oxidative phosphorylation". There are specific enzymes, "biomolecular nanoreactors", organized in the respiratory electron transport chains (ETC). Normal function of the ETC enzymes, be it mitochondrial nanoreactors in eukaryotic cells of animals or similar nanoreactors of bacteria cells, is in the transport of electrons, one by one, from the electron donor molecules to the end enzyme, cytochrome oxidase, from which the electrons are transferred to molecules of oxygen with two electron reduction of oxygen into water. Free energy released during the electron transport is used by the specific enzyme ATP-synthase, for synthesis of ATP from adenosine 5'-diphosphate (ADP) and inorganic phosphate.

In the experiments with mitochondria isolated from the rat hearts, it has been revealed that the rate of oxidative phosphorylation with magnetic ^{25}Mg was 2-3 times higher than that with nonmagnetic ^{24}Mg and ^{26}Mg while no difference was found between the nonmagnetic magnesium nuclei. It was also revealed that activity of phosphocreatine kinase and phosphoglycerate kinase, for which ions of Mg^{2+} serve obligatory cofactors, was essentially higher with magnetic ^{25}Mg than with ^{24}Mg and ^{26}Mg. Again, no difference in efficiency between the nonmagnetic magnesium nuclei was found in these experiments. Thus, there have been the very first evidences of MIE in biochemical reactions in vitro. Furthermore, the same research group has discovered MIE of calcium. The activity of phosphocreatine kinase with Ca^{2+} ions of magnetic nuclei ^{43}Ca was found to be twice higher than with Ca^{2+} ions of nonmagnetic nuclei ^{40}Ca.

Factual evidence of MIE, on its own, indicates that there is a spin-selective "bottle-neck" of the process under investigation. The hypothetic mechanism of the acceleration of oxidative phosphorylation by the nuclear spin of ^{25}Mg suggested by this group is as follows. Namely, they suggested a reversible transfer of electron density in the active center of ATPsynthase from the terminal anion phosphate group of ADP to Mg^{2+}-cation. It produces a virtual ion-radical pair, Mg^{+}-adenosine phosphate radical in the singlet spin state.

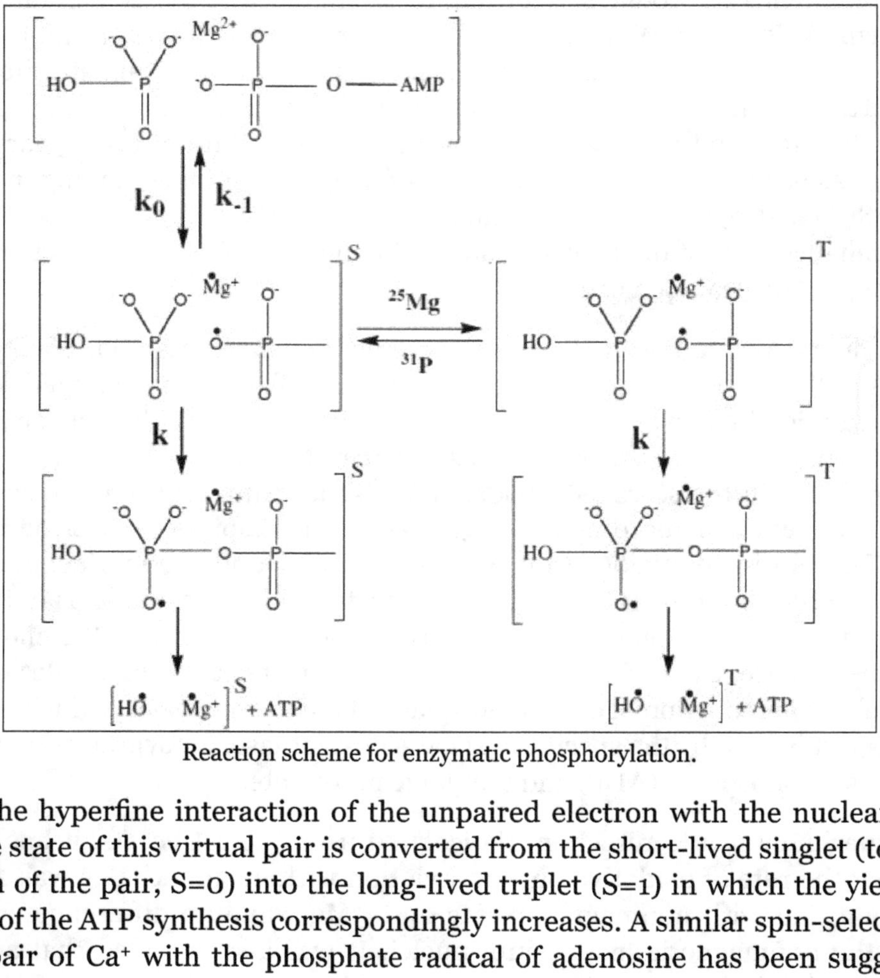

Reaction scheme for enzymatic phosphorylation.

Due to the hyperfine interaction of the unpaired electron with the nuclear spin of ^{25}Mg, the state of this virtual pair is converted from the short-lived singlet (total electron spin of the pair, S=0) into the long-lived triplet (S=1) in which the yield of the reaction of the ATP synthesis correspondingly increases. A similar spin-selective ion-radical pair of Ca^{+} with the phosphate radical of adenosine has been suggested to explain MIE of ^{43}Ca.

It should be mentioned that the hypothesis about a possible key-role of such virtual ionradical pairs in oxidative phosphorylation and energy transformation processes has long been stated. Within the context of modern bioenergetics, which postulated a proton electrochemical gradient across the mitochondrial membrane as the energy-rich intermediate of oxidative phosphorylation in the "molecular motors", another plausible explanation for the MIE can be proposed. It is reasonably to suggest that the proton electrochemical gradient poses conformational pressure in the catalytic center of ATP-synthase, generating electronic-conformational excitation in the ADP-Mg^{2+} complex. This would essentially increase reactivity of the adenine base of ADP to phosphate. The nuclear spin of ^{25}Mg (or ^{43}Ca) can provoke the transition of the exited ADP-Mg^{2+} complex from the singlet state to the triplet state the lifespan of which is longer, thereby providing more time for the reaction of the ATP synthesis from ADP and Pi. Thus, the detailed mechanism of the magnetic-isotope catalysis in bioenergetics remains to be cleared.

In Situ: Magnetic-isotope Catalysis in Living Cells

There is a great variety of chemical elements in biomolecular nanoreactors of living cells. Certain of them are only represented by magnetic isotopes, among them – hydrogen, nitrogen, sodium, phosphorus, potassium, manganese and so on. However, there are chemical elements which have both kinds of stable isotopes, nonmagnetic and magnetic ones, among them – carbon, oxygen, magnesium, calcium, iron, zinc and others. Correspondingly, these are the elements which are required to search for magnetic-isotope effects in living cells.

In this regard, magnesium is of particular interest. There are three stable isotopes of magnesium, ^{24}Mg, ^{25}Mg and ^{26}Mg with natural abundance about 79, 10 and 11 %. Among them, only ^{25}Mg has the nuclear spin (I=5/2) that produces the magnetic field. Two other isotopes are spinless (I=0) and, hence, produce no magnetic fields. As the most abundant intracellular divalent cation, Mg^{2+} is essential to regulate numerous cellular functions and enzymes. Ions of Mg^{2+} serve as obligatory cofactors in catalytic centers of many enzymes including ATP-synthase as the primary producer of ATP in mitochondria, chloroplasts, bacteria and archaea. Moreover, a novel role for Mg^{2+} as an intracellular second messenger has been recently discovered. Besides, the difference in masses between the isotopes of magnesium is much less, in percentage term, than that for the isotopes of carbon, for example, thereby minimizing the classical mass-isotope effect.

Stable isotopes of magnesium, namely nutrient solutions highly enriched with ^{25}Mg or ^{26}Mg, have been used for many years as in vivo tracers to determine magnesium absorption in human subjects, animals and plants models. It is reasonable that the problem of possible beneficial effects of the magnetic isotope was not posed in the cited papers. There have been attempts to use the non-radioactive isotopes to make the oxidative biomolecules more stable against free-radical oxidation. It was found that deuterated polyunsaturated fatty acids protect yeast cells against the toxic effects of lipid autoxidation products. Both natural isotopes of hydrogen, ^{1}H and ^{2}H (D), are magnetic ones, but they have the twofold difference in masses. Consequently, the observed protection should be ascribed to the mass-isotope effect instead of the magnetic isotopy. In the in vitro study of cleavage of deuterated DNA by the hydroxyl radical, the value of the kinetic effect was found to be close to 2, just the mass-ratio of hydrogen and deuterium. Similar mass-isotopic effects are, presumably, anticipated for stable isotopes of nitrogen ^{15}N versus ^{14}N and for carbon ^{13}C that is 8% heavier than ^{12}C.

Recently, the smart methods of labeling of nematodes Caenorhabditis elegans, which are commonly used in gerontology, with ^{13}C and ^{15}N have been developed by feeding the worms with heavy isotope–labeled Escherichia coli. However, with regards to the ideas of using the stable isotopes and their mass-isotope effects to stabilize cells against free-radical oxidation and, thereby, improve the living conditions and even extend the lifespan, there are doubts if high amounts of ^{13}C and ^{15}N in the numerous proteins, DNA, RNA, and other molecules of living cells will have, eventually, beneficial effects rather than harmful ones.

In searching beneficial isotope effects, including possible distinctions between the effects of magnetic and nonmagnetic isotopes, the very first step should be preparation of the cells enriched with different isotopes of magnesium. With this aim, we used bacteria E. coli, the commonly accepted cell model, and the growth media of the identical chemical composition with one exception, that they were supplemented with different isotopes of magnesium, magnetic ^{25}Mg and nonmagnetic ^{24}Mg or ^{26}Mg, as $^{25}MgSO_4$, $^{24}MgSO_4$ or $^{26}MgSO_4$. The oxides of magnesium, ^{24}MgO, ^{25}MgO and ^{26}MgO, with isotope enrichment 99.9, 98.8 and 97.7 atom percent, correspondingly, were purchased from RosAtom, Russia. $^{24}MgSO_4$, $^{25}MgSO_4$ and $^{26}MgSO_4$ were prepared from the relevant oxides by using a standard acidic treatment with analytically pure sulphuric acid.

The pioneering studies have been done in Institute of Problems of Chemical Physics, RAS. Bacteria E. coli, strain BB, were cultivated in accordance with the standard design on the artificial liquid minimal M9-media, composed from 8 g of glucose, 2 g of NH4Cl, 12 g of Na2HPO4, 6 g of K2HPO4, 1 g of NaCl in 750 ml of distilled water and 250 ml of tap water as the source of microelements. The growth media were supplemented with the isotopes of magnesium so that the final concentration of $^{24}MgSO_4$ or $^{25}MgSO_4$ was 2.2 mM per liter of the media. Cells were grown aerobically at 37 °C with shaking, harvested at the late growth ("stationary" phase at OD600 of about 0.5), and viability of the cells was tested as their ability to form colonies (colony-forming units, CFU) on the solid nutrient BCP agar using standard Petri's dishes.

The experimental data are presented in figure. The striking effect of the magnetic isotope, ^{25}Mg, has been detected when tallying up the colony forming units. The standard nutrient agar contains all components necessary for normal growth of cells, including magnesium. Nevertheless, the amount of CFU formed by the bacteria, which were previously grown on magnetic ^{25}Mg, has turned out to be about 40 percent higher in comparison with the bacteria, which were previously grown on nonmagnetic ^{24}Mg. Thus, the cells which have been previously enriched with the magnetic isotope of magnesium demonstrate the essentially higher viability in comparison to the cells enriched with the nonmagnetic isotope.

Another striking effect of the magnetic isotope of magnesium has been detected when measuring activity of superoxide dismutase (SOD), the main antioxidant enzyme of the cells. The cells enriched with ^{25}Mg demonstrate the reduced activity of SOD, about 40 percent, when compared to the cells enriched with ^{24}Mg. It is generally known that E. coli growing aerobically contain MnSOD and FeSOD. Inasmuch as the total SOD activity is normally adjusted to the intracellular level of O_2 •—, the reduced level of SOD in the cells can be considered as evidence for lower production of O_2 •— as the failure by-product of cell respiration.

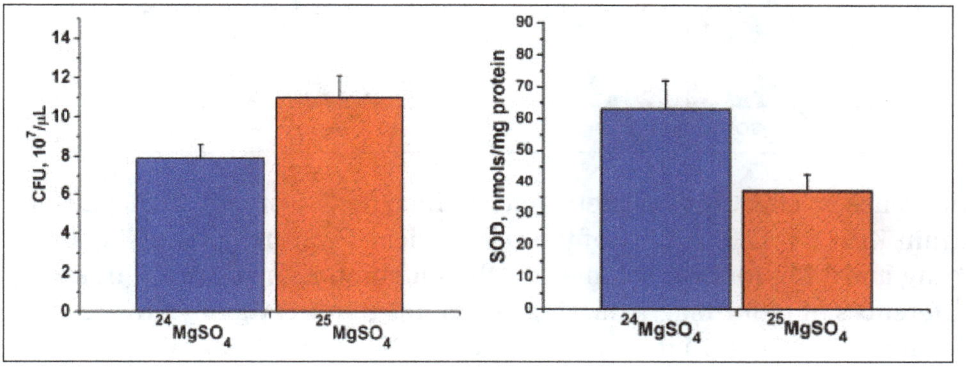

The difference in effects of magnetic and nonmagnetic isotopes of magnesium on E. coli, strain BB. Left: ability of the cells enriched with magnetic ^{25}Mg or nonmagnetic ^{24}Mg to form colonies on the nutrition agar. Right: activity of superoxide dismutase in the cells enriched with magnetic ^{25}Mg or nonmagnetic ^{24}Mg. Data are indicated as m ± SD, N = 3. The difference of the mean values (m) for magnetic ^{25}Mg vs. nonmagnetic ^{24}Mg is statistically significant at $P \leq 0.01$.

Thereafter, these experiments have been replicated in cooperation with the microbiologists of Orenburg State University using another strain of E. coli, K12TG1. After cultivation for 24 h in the artificial liquid minimal M9-medium without magnesium (and without tap water), the cells were suspended in the fresh M9-medium supplemented with different isotopes of magnesium as 0.26 g of $^{24}MgSO_4$, $^{25}MgSO_4$ or $^{26}MgSO_4$ per liter, and grown aerobically at 37 °C. To obtain reproducible results, three parallel experiments with each kind of the isotopes have been simultaneously performed, i.e. bacteria supplied with ^{24}Mg, or ^{25}Mg, or ^{26}Mg, three samples each, were simultaneously tested in the same experimental succession. After reaching the stationary phase with OD620 of about 0.5, the cells were tested for their ability to form colonies after inoculation on the agar plate's surfaces.

From the experimental results, shown on figure., one can see that the amount of CFU formed by the bacteria, previously grown on magnetic ^{25}Mg, has turned out to be almost twice higher in comparison with the bacteria which were previously grown on nonmagnetic ^{24}Mg and ^{26}Mg. Noteworthy, there has been no significant difference between nonmagnetic ^{24}Mg and ^{26}Mg in their effects on CFU. It gives evidence that there is the magnetic isotope effect of ^{25}Mg rather than a classical mass-dependent isotope effect of the magnesium isotopes.

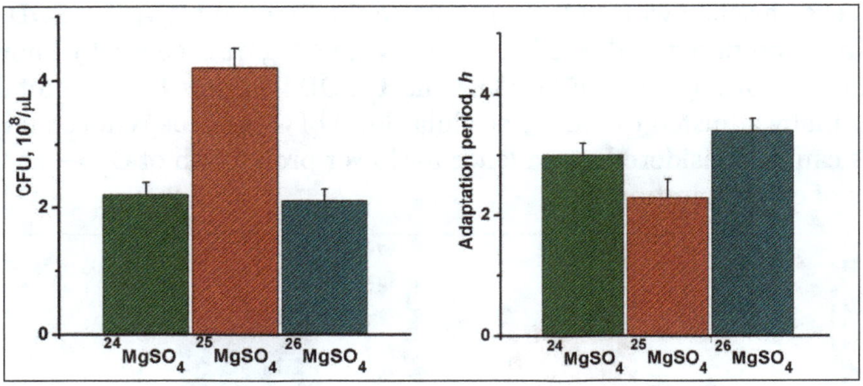

The difference in effects of magnetic and nonmagnetic isotopes of magnesium on E. coli, strain K12TG1. Left: Ability of the cells enriched with magnetic ^{25}Mg or nonmagnetic ^{24}Mg and ^{26}Mg to form colonies on the nutrition agar. Data are m ± s.d., N =3. The differences of m for magnetic ^{25}Mg vs. nonmagnetic ^{24}Mg or ^{26}Mg are statistically significant at $P < 0.005$. Right: Length of adaptation period (lag-phase) on the liquid media supplemented with magnetic ^{25}Mg or nonmagnetic ^{24}Mg and ^{26}Mg. Data are m ± s.d., N =3. The differences of m for ^{25}Mg vs. ^{24}Mg or ^{26}Mg are statistically significant at $P \leq 0.02$.

In addition, kinetics of the cell growth was hourly monitored in the cited experiments, using optical density measurements at 620 nm with 96-cavity micro-plate reader "Uniplan" (Picon, Russia). The kinetic curves of cell biomass growth were typical for the bacterial growth involving slow adaptation period ("lag-phase") followed by exponential growth ("logphase") when the cell mass quickly doubled and the stationary phase when the cell growth completed because of lack of the substrates (primarily, glucose). The striking observation for these experiments was that length of the lag-phase has turned out to be essentially shorter in the case when the cells were transferred on the media with magnetic isotope of ^{25}Mg in comparison with nonmagnetic ^{24}Mg and ^{26}Mg. Again, the nonmagnetic isotopes, ^{24}Mg and ^{26}Mg, were not differentiated by their effects on the cell growth. It means that adaptation of cells transferred from the pre-incubation media is limited by some enzymatic processes, efficiency of which is increased by magnetic nuclei of ^{25}Mg.

Magnetic isotope of ^{25}Mg more effectively performs the Mg^{2+} cofactor function for oxidative phosphorylation in the isolated mitochondria in comparison to ^{24}Mg or ^{26}Mg. Energetic demands of every operation in prokaryotic cells of bacteria, as well as in eukaryotic cells of animals, are met by molecules of ATP. Hence, ATP as the main source of energy in living cells is most likely to be the limiting substrate for adaptation metabolism in the lag-phase. However, for the exponential phase of the cell growth, we have not found any significant differences in the rates of cell growth between magnetic ^{25}Mg and nonmagnetic ^{24}Mg and ^{26}Mg. The time of doubling of the cell mass was approximately the same, regardless of the type of magnesium isotopes. This obviously suggests a different "bottle-neck" of metabolism in the exponential phase of growth with other, than ATP, limiting substrate and another limiting reaction independent upon the nuclear spin of magnesium.

Of special interest are searches for magnetic-isotope effects in the processes of recovery of cells from radiation injuries. The reason is that any factor, capable to influence on efficiency and reliability of cell nanoreactors, shows up itself most vividly under drastic conditions of post-radiation recovery.

We undertook the investigation of effects of magnetic and nonmagnetic isotopes of magnesium on post-radiation recovery of S. cerevisiae. The yeast cells were cultivated on the standard nutrient liquid media M_3 supplemented with $^{24}MgSO_4$ or $^{25}MgSO_4$. After three days of the cultivation under aerobic conditions at 30 °C, the cells were washed from the nutrient liquid and suspended in nutrient-free ("fasting") media, i.e., sterile phosphate buffer, pH 7.0. Then, the cells were irradiated by the short-wave ultraviolet light (λ ≈240 nm, the dose ≈190 J/m²), whereupon they were left in the nutrient-free water at 30 °C (with shaking) to study kinetics of the post-radiation recovery of the cells. For this kinetics study, the aliquots of the cells were periodically seeded on the standard nutrient (Petri dishes) and the cell survival was monitored by the standard CFU technique.

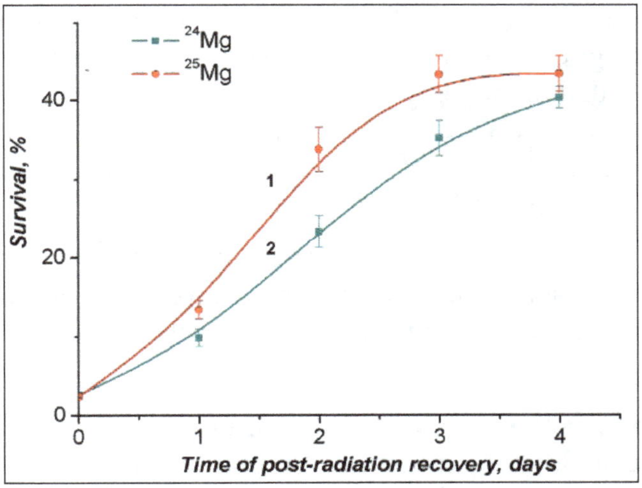

The difference in effects of magnetic and nonmagnetic isotopes of magnesium on post-radiation recovery of S. cerevisiae, diploid yeast cells. The cells enriched with magnetic ^{25}Mg or nonmagnetic ^{24}Mg were irradiated by the short-wave UV-C light (λ = 240-260). Survival of cells was estimated as their ability to form colonies on the nutrition agar: 1 - recovery of the cells enriched with ^{25}Mg; 2 – recovery of the cells enriched with ^{24}Mg.

The survival of cells transferred to agar immediately after irradiation was not more than a few percent. In this case the injured genetic structures in most of the cells could not be repaired before the onset of mitosis, and nonviable daughter cells were produced. Incubation in nutrient-free water, in which the cells do not divide, provides sufficient time for repair processes and leads to a corresponding increase in survival. From the kinetics curves represented on this figure we notice that the cells enriched with magnetic isotope of magnesium, ^{25}Mg, are recovered essentially more effectively than the cell enriched with the nonmagnetic ^{24}Mg.

It has been known that kinetics of recovery of yeast cells from radiation injuries may be represented by a function representing the reduction of the effective radiation dose, Deff, with time:

$$D_{eff}(t) = D_o\left[k + (1-k)\exp(-\beta t)\right]$$

In this model of A. Novick and L. Szilard, Do is the radiation dose, t is time of post-radiation recovery in the nutrient-free water, β is the recovery rate constant, and κ is the fraction of irreversible injuries.

Table represents values of the kinetics parameters resulting from these experiments. While the fraction of irreparable injuries remained almost the same, the value of the rate constant β of the post-radiation recovery was twice higher for the cells enriched with ^{25}Mg than for the cells enriched with ^{24}Mg. This is decisive evidence that the magnetic isotope of magnesium essentially more effectively promotes the recovery of cells from radiation damages. Thus, we have, for the first time, documented the magnetic-isotope effect in radiation biology.

	β, h^{-1}	k
^{24}Mg	0.032 ± 0.003	0.70 ± 0.14
^{25}Mg	0.058 ± 0.004*	0.61 ± 0.12

Effect of magnetic ^{25}Mg isotope on postradiation recovery of S. cerevisiae, diploid yeast cells, after short wave UV irradiation. *Difference between the means is significant at P = 0.02.

One might suggest that the observed effects in our experiments with bacteria and yeast cells were caused by different levels of impurities in the growth media complemented with different isotopes of magnesium. However, it could hardly be the case. First, according to the mass-spectrometry data, amounts of contaminant elements in the stock solutions of the isotopes did not exceed 20-30 ppm, be it sulphate of ^{24}Mg, ^{25}Mg, or ^{26}Mg. Second, amounts of the contaminants that were administered in the liquid growth media with glucose and other basic components have significantly exceeded amounts of the same contaminants administered with much less additions of the isotope stock solutions. Besides, the impurities that were administered into the growth media from the basic components, as well as the element contents of the solid nutrient agar media, were obviously the same in all experiments, independently of the magnesium isotopes. Hence, one can disregard impurities as a possible reason of higher efficiency of magnetic ^{25}Mg than that of nonmagnetic ^{24}Mg and ^{26}Mg. It is apparent that the cells in the above cited experiments perceive the difference between magnetic and non-magnetic isotopes of magnesium, i.e., they perceive the nuclear spin's magnetic field of ^{25}Mg.

Future Prospect: Nuclear Spin as Factor of Reliability in Cell Nanoreactors

Thus, our data have documented, for the first time, the magnetic-isotope effect of magnesium-25 in vivo. Factual evidence of MIE, on its own, indicates that there is a spinselective "bottle-neck" of the process under investigation. Ions of Mg^{2+} perform not only the cofactor functions in synthesis and hydrolysis of ATP. In addition, they have the impact on the structure-functional properties of RNA, RNA-polymerase, ribonuclease, and so on. Besides, there are the specialized proteins which regulate homeostasis and transport of Mg^{2+} in living cells. Moreover, ions of Mg^{2+} may work as the intracellular second messengers. Up to date, however, there have not been findings of MIE except for the enzyme synthesis of ATP in the above cited papers. The similar MIE of ^{25}Mg is assumed to be in our experiments. Indeed, adaptation of cells to novel growth conditions requires a large variety of stress proteins to be synthesized and ATP, as the main source of energy in microbial cells, is most likely to be the limiting substrate for the adaptation metabolic processes. Similarly, a large variety of biosynthesis is required for recovery of cells from radiation injuries. It is reasonable to suggest that the kinetics of post-radiation recovery is also limited by spin-selective synthesis of ATP as the "bottle-neck". The post-radiation recovery proceeds with higher rate when the cell nanoreactors run on the magnetic isotope of magnesium, because the nuclear spin of ^{25}Mg catalyzes the ATP synthesis, hereby supplying the cells with more amount of ATP.

The lower level of superoxide dismutase (SOD) activity in the E. coli cells enriched with ^{25}Mg, by about 40 per cent when compared to the cells that were grown on nonmagnetic ^{24}Mg, can be also flow from the beneficial effect of the magnetic isotope on the ATP synthesis. The cell nanoreactors of oxidative phosphorylation have very ancient evolutionary origin and, hence, seem to be ones of the most reliable biomolecular machines. But yet their reliability ("robustness") characteristics are not perfect because these molecular machines experience conformational fluctuations. It is well known that normal elementary acts of electron transfer on the electron transport chains, be it mitochondria or prokaryotes, alternate with random malfunctions when an electron, rather than waits for transport to the next enzyme of the electron-transport chain, goes directly to an adjacent oxygen molecule. Such an electron leakage results in production of $O_2^{\bullet-}$. Chemical products of $O_2^{\bullet-}$, the so-called reactive oxygen species (ROS), are toxic and initiate free-radical damages in the biopolymer nanoreactors. SOD catalyzes the reaction of dismutation of $O_2^{\bullet-}$ into hydrogen peroxide (H_2O_2) and oxygen, thus protecting cell structures from $O_2^{\bullet-}$ and its toxic chemical products. It makes its defense "job" in cooperation with two other specific enzymes, catalase and glutathione peroxidase, which catalyze decomposition of H_2O_2 into nontoxic reagents, namely H_2O and O_2.

As a rule, the level of the SOD activity is adjusted to the intracellular level of $O_2^{\bullet-}$. If SOD activity decreases or increases, it normally reflects the relevant decrease or increase in the production of $O_2^{\bullet-}$ as faulty by-products of the electron-transport nanoreactors of oxidative phosphorylation. Hence, the lower level of SOD activity testifies the lower production of $O_2^{\bullet-}$ in the case when the cells are supplied with the magnetic isotope. As cited above, oxidative phosphorylation of ADP proceeds faster with ^{25}Mg by comparison with ^{24}Mg. Since ^{25}Mg is more effective cofactor of oxidative phosphorylation, it transpires that the ATP synthase operates faster with the magnetic magnesium nucleus by comparison with the nonmagnetic ones. Meanwhile, under normal coupled conditions, the electron transport is subjected to the "backpressure" of the respiration-generated transmembrane electrochemical H^+ - gradient. The partial dissipating of this gradient via the acceleration of ADP phosphorylation has been shown to result in the decreased production of $O_2^{\bullet-}$. Hence, the higher rate of oxidative phosphorylation with magnetic ^{25}Mg, in comparison with nonmagnetic ^{24}Mg, through the reduction of the H^+-gradient "backpressure" should decrease the false electron leakage onto oxygen, thereby reducing the yield of free radicals $O_2^{\bullet-}$ as by-products of the electron transport.

From the point of view of chemical kinetics, with decrease in the rate of oxidative phosphorylation, there is the retardation of electron transport in the sites of the electrontransport chains which are coupled with phosphorylation of ADP. Inasmuch as the input of electron-transport nanoreactors becomes overflowed with electrons ("electron-transport jam"), the probability of electron leakage on oxygen increases. The more is acceleration of the ATP synthesis, the less is probability of the "jam". Hence, the yield of $O_2^{\bullet-}$as the byproducts of electron transport is bound to be much lower with ^{25}Mg-ADP by comparison with ^{24}Mg-ADP or ^{26}Mg-ADP.

Thus, with magnetic ^{25}Mg, the biopolymer nanoreactors of oxidative phosphorylation operate not only more effective but more reliable too, in comparison with their operation on non-magnetic isotopes ^{24}Mg and ^{26}Mg. Downgrading production of $O_2^{\bullet-}$, the magnetic isotope of magnesium produces, actually, a beneficial preventive antioxidant effect. This antioxidant effect of the nuclear spin, that ^{25}Mg favors the less production of reactive oxygen species, should obviously increase longevity of the electron-transport nanoreactors. Therefore, it can reveal itself in nature as the kinetic nuclear-spin selection of the favorable isotope, namely, the kinetic isotope enrichment with magnetic ^{25}Mg in the processes of recycling and regeneration of the electron-transport nanoreactors. For example, one can expect for the enrichment with the favorable magnetic ^{25}Mg in recycling and regeneration of cell mitochondria with aging of the cells.

Contrary, in the case of photosynthetic nanoreactors, one can predict undesirable prooxidant effect of the magnetic nucleus of ^{25}Mg. Indeed, it is known that the function of the vast majority of chlorophyll molecules (Chl), as the derivatives of the magnesiumprotoporphyrin complexes, is to absorb light energy and transfer it to the specific energy sinks, the so-called reaction centers of the photosynthetic nanoreactors. While performing this energy-transfer function, the light-exited Chl molecules are in the singlet state (1Chl*, electron spin S=0). However, there is probability of the radiationless relaxation into the triplet state (3Chl, S=1) followed by formation of singlet oxygen, $1O_2$, the molecules of which are substantially more reactive by comparison with usual triplet O_2 molecules and, thereafter, produce oxidative damages. As nuclear spin of ^{25}Mg can catalyze the conversion of 1Chl* into the triplet 3Chl, one can expect for the higher yield of $1O_2$ and, thereafter, more photodynamic damages in the cells with the chlorophyll molecules containing ^{25}Mg instead of the spinless ^{24}Mg or ^{26}Mg. Correspondingly, it is beyond reason to hope for selection of the magnetic Mg, in the case of algae or green plants. Besides, the functional disadvantage of ^{25}Mg should be followed by increased synthesis of carotenoids and other natural antioxidants. Indeed, measurements of magnesium isotopic composition of the chlorophylls extracted from cyanobacteria and similar analysis of the chlorophyll forms in the leaves of English Ivy (Hedera helix L.) have revealed the isotope distribution following usual classical mass-isotope effect with no evidence for depletion or enrichment of ^{25}Mg.

Apart from magnesium, there are many other elements which have both kind of stable isotopes, nonmagnetic and magnetic ones, amongst them – carbon, oxygen, calcium, iron and zinc. In passing, nuclear spin of ^{17}O should lift the spin ban over the reaction of the mitochondrial ubisemiquinone radicals with oxygen, thereby catalyzing formation of $O_2^{\bullet-}$. As a result, one can expect for more reactive oxygen radicals and more free-radical damages due to ^{17}O in comparison with nonmagnetic ^{16}O and ^{18}O. The pro-oxidant action of ^{17}O can reveal itself as the selective enrichment of free-radical peroxidation products with this unfavorable magnetic isotope as compared with ordinary metabolites.

OPTICAL FIBER GRATINGS IN BIOMEDICINE

Optical fiber is a flexible, transparent waveguide or "light pipe" to transmit light; the latter, in its various forms and facets has caught the attention of humanity since prehistoric times. The ancient civilization used it as fire signals to communicate and later as a therapeutic and preventive tool for better health. In the modern era, the idea that light can be used for communication combined with the phenomenon of total internal reflection, gave rise to the concept of medium for light transmission. As a consequence, by the end of 19th century glass rods as illuminators were realized. Optical fibers were the next step as they are basically glass rods stretched very thin to become long and flexible. Gradual technological advances from 1920s when use of fiber for light transmission was first proposed, to 1980s resulted in glass fibers as the most ideal communication medium for enormous amount of data with lowest possible attenuation. Their inherent properties such as small size (have standard thickness of ~0.250 mm that can be less than that of surgical suture), biocompatibility, non-toxicity, chemical inertness and remote monitoring capability, make them quite lucrative for usage in the biomedical area. These fibers thus have diverse applications ranging from illumination to imaging, from phototherapy to precise surgery, from monitoring complex biomechanical dynamics to wearable smart sensors. In fact, after their first practical application in flexible endoscopes reported by Basil Hirschowitz in 1957 for illuminating and imaging internal organs of human body, the optical fibers have come a long way as sensors for various physiological parameters as well.

Fiber Gratings

Fiber gratings are one of the simple intrinsic fiber devices that can reflect, filter or disperse light passing through them, suitable not only for communication applications but also finding their foothold as a fascinating sensor element in diverse areas. Figure shows the schematic representation of a typical single mode fiber with fiber grating inscribed in its core.

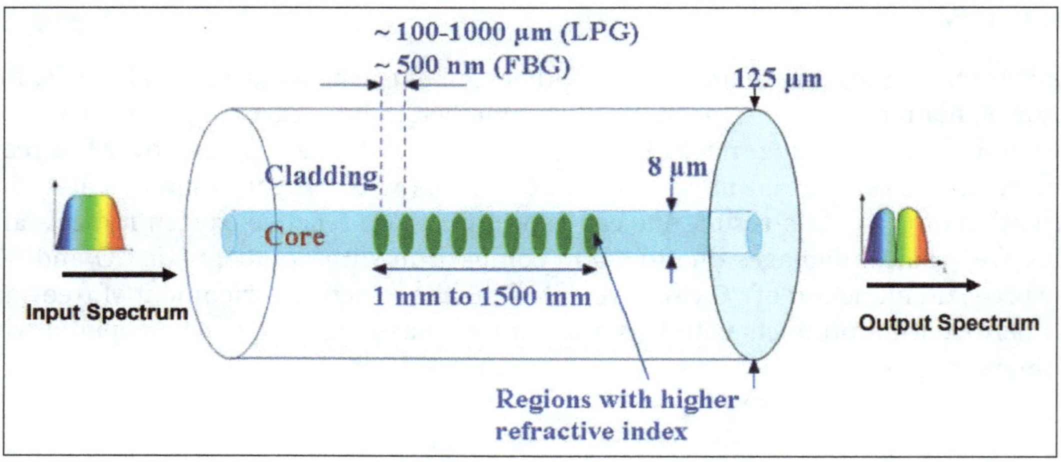

A typical single mode fiber with fiber grating.

Special Features

Optical fibers are long and flexible waveguides generally made of fused silica (amorphous silicon dioxide, SiO_2) the most abundant mineral found in the earth crust. Though plastic optical fibers made of polymers are also quite practical and inexpensive for some applications, our focus here is on silica fibers which are the backbone of optical communication today. Their potential as efficient sensor elements was recognized simultaneously because of the unique advantages they offer, such as:

- Small size and geometric versatility: with diameter in the range of 125 -500 micrometers, fiber sensors can be configured in arbitrary shapes and offer a great choice in space-restricted or hard to reach environments.

- Common technology base for multiparameter sensing: a single fiber can be used to sense various physical perturbations (acoustic, magnetic, temperature, strain, pressure, rotation, etc.)
- Compatibility with communication fiber facilitates telemetry.
- Absence of crosstalk between close fibers suggests that different sensors can be housed in close vicinity e.g. a catheter.
- A single electro-optic unit can be utilized for all the sensors, naturally with an appropriate illumination, detection and signal-processing scheme.
- Optical fibers have very low attenuation and thus long fiber links can be used to remotely monitor and control the sensing parts without disturbing the patient while keeping all the electronics away.
- Immunity to electromagnetic/radiofrequency interference and chemically inert nature: due to their dielectric construction, they can be used in high voltage, electrically noisy, high magnetic field, high temperature, corrosive, or other harsh environments

Along with these, fiber gratings offer some added features unique to them like selfreferencing as the information is wavelength encoded and ease of multiplexing, facilitating distributed sensing, that make them more valuable as sensors. Their potential as strain and temperature sensors array in various concrete, metal and composite structures has been established much earlier and they have been implemented successfully for structural health monitoring (SHM) in the arena of civil and aerospace, oil & gas exploration wells, power systems monitoring, etc.

Relevance in Biomedicine

Because the fiber gratings are made of dielectric glass, they are inherently immune to electromagnetic interference (EMI) and can be safely used without any electrical or chemical obstruction in a clinical setup. Their miniature size, flexibility, and lightness allow easy insertion in catheters/needles making possible localized measurements inside blood and tissues. Also, multiple sensors can be accommodated on a single fiber, working independently of the other. Silica is chemically inert and fulfills the biocompatibility criterion. As fibers are intrinsically safe for the patient i.e. their use produces no immunity response from human defense system, they can be used for in- vivo measurements and can be left in their position for repeated or continuous monitoring.

Although all these qualities are quite exciting there are some issues coming into the way of fiber grating technology to become accepted in healthcare systems that remain to be dealt with. For example, howsoever small or non-intrusive these sensors are for the patient, their read out units need to be connected through a fiber link which is not so practical in the modern world of wireless technology. With the emergence of smaller and faster interrogators one possible way out can be the development of very small wearable interrogator system with no external fiber links. Another issue is information extraction process that needs to be standardized for each application. Fiber gratings provide information in terms of wavelength shift that can be due to changes in various external parameters e.g. strain, temperature, pressure, refractive index (RI) etc. when the sensor is being used for one specific measurand (strain for example) it should be unaffected by all other parameters (temperature, RI etc.) to minimize error. This can be achieved by using optimized packaging/transducer. Also, if one sensor is being used for multiparameter sensing, it will require customized interrogation system and software tool to discriminate the effect of each parameter. Practical solutions of all these issues require multidisciplinary approach with a synergy between various experts e.g. doctors, physicists and engineers.

Working Principle

Optical fiber consists of an inner dielectric core surrounded by cladding of another dielectric material of slightly lower refractive index than that of core. In the simplest case, the waveguide effect is often explained as resulting from repeated total internal reflection (TIR) of light rays at the core–cladding interface. The fiber can then guide all light impinging the input face under certain conditions. Standard fibers have core and cladding refractive indices constant along the fiber length, but if a periodic modulation of refractive index is created deliberately along the fiber core it results in formation of fiber gratings.

The periodic variation of refractive index in the fiber core is created mostly by its exposure to Ultraviolet (UV) radiation through a proper mask. These gratings are categorized as short period or fiber Bragg gratings (FBGs) and long period gratings (LPGs). In an FBG a narrow band of wavelength is reflected and there is a corresponding drop of intensity in the transmitted spectrum while LPGs work as wavelength dependant loss elements exhibiting multiple loss resonance bands in the transmitted spectrum. Though their ability to filter certain wavelengths was a major attraction in the field of telecommunication, their potential as sensing devices was recognized simultaneously. This is because the filtered wavelength is a function of its effective refractive index as well as the period of the grating and any variation in these parameters results in shift of wavelength which can be detected easily using an optical spectrum analyzer (OSA) or an interrogator. Apart from basic FBG and LPGs there are some other distinct grating types that are formed when deliberate nonuniformity in the refractive index profile is introduced. Blazed or tilted FBGs are resulted when grating planes are at an angle with the fiber axis and are used mostly for mode conversion. A tilted grating can be designed in a way that its' core mode is coupled with some of the cladding modes so that the Bragg Wavelengh becomes a function of ambient refractive index and thus can be used for refractive index sensing. Another type is chirped FBG in which the grating period is aperiodic having a monotonic increase/decrease or nonuniformity longitudinally. In this type of gratings each point has different Bragg wavelength and hence its spectrum can be used to monitor a parameter profile by distributed sensing.

In an FBG, the guided light is scattered by each interface of different refractive index regions in the core and for a wavelength which satisfies Bragg Condition, the scattered light adds up constructively resulting in back reflection with a central wavelength (λ_B) given by:

$$\lambda_B = 2n\Lambda \; [\text{Bragg Condition}]$$

Where Λ is the pitch or periodicity of the grating, n is the effective refractive index of the core and λ_B is the Bragg wavelength. Therefore, when light from a broadband source is launched in this FBG the spectral component defined by above equation is missing from the transmitted spectrum. Bragg wavelength is shifted if the effective refractive index, the grating periodicity or both are changed due to some perturbation; in fact both these parameters are directly influenced by strain and ambient temperature with the associated wavelength shift given as:

$$\Delta\lambda_B = 2\left[\Lambda\frac{\partial n}{\partial l} + n\frac{\partial \Lambda}{\partial l}\right]\Delta l + 2\left[\Lambda\frac{\partial n}{\partial T} + n\frac{\partial \Lambda}{\partial T}\right]\Delta T$$

Where Δl is change in grating length due to strain and ΔT is change in ambient temperature. The first term on the RHS gives strain dependence while the second term indicates temperature dependence of the Bragg wavelength and an FBG sensor works by monitoring Bragg wavelength shift with one or both of these parameters.

LPGs couple fundamental guided core mode to different cladding modes. The loss resonance wavelength at which this coupling takes place satisfies the phase matching condition i.e.:

$$\lambda_i = \left[n_{eff}^{co} - n_{ieff}^{cl}\right]\Lambda$$

Where co n_{eff}^{co} and n_{ieff}^{cl} are the effective refractive indices of the fundamental core mode and i th cladding mode respectively and Λ is the period of the LPG. Since effective index of a cladding mode is dependent upon the refractive index of the surrounding medium, any change in the latter alters the loss resonance wavelength. The influence of refractive index of the surrounding medium on the LPG wavelength is expressed by the following equation:

$$\frac{d\lambda_i}{dn_{sur}} = \frac{d\lambda_i}{dn_{ieff}^{cl}} \cdot \frac{dn_{ieff}^{cl}}{dn_{sur}}$$

Where, n_{sur} is the refractive index of the surrounding medium.

Apart from the shift in loss resonance wavelengths, the variation in nsur is also reflected as variation in intensity of the loss resonance peak, defined by the overlap integral I given by equation below,

$$I = \frac{\int_r \int_\phi \psi_{core} \psi^*_{clad} r dr d\phi}{\sqrt{\int_r \int_\phi \psi_{core} \psi^*_{clad} r dr d\phi} \sqrt{\int_r \int_\phi \psi_{core} \psi^*_{clad} r dr d\phi}}$$

Where,ψ s are the electromagnetic field components of the two coupling modes, r and Φ are radial and azimuthal co-ordinates respectively. Obviously, any change in cladding field distribution will affect the coupling strength. When the refractive index of the surrounding is varied, it will alter the cladding field distribution and hence the overlap integral. Since effective refractive index of a cladding mode and their coupling efficiency with the fundamental mode is dependent upon the refractive index of the surrounding medium, any change in the latter is easily detected using LPGs. This is the basic principle of an LPG based refractive index and concentration sensor.

Fabrication Technology

Direct inscription of submicron periodic pattern in optical fibers puts severe constraints in terms of stability and precision on the grating fabrication techniques. There are only a few methods, namely, the interferometric, the phase mask, and the point-by-point techniques that give consistently good quality gratings.

In our laboratory, fiber gratings are produced by exposing core of a photosensitive fibe to intense UV light from a KrF (Krypton Fluoride) excimer laser at 248nm wavelength. The primary acrylate coating is first removed from the fiber section that is to be exposed to UV light, using a thermo-mechanical stripper to avoid mechanical degradation. This uncoated fiber is further loaded onto the fiber mounting stage and an appropriate tension is applied to keep it straight, using load cell. After that the fiber is first made to approach close proximity of the mask within 0. 5mm and vertical & horizontal alignments are performed by continuously monitoring on a computer screen. To inscribe FBGs the light from the KrF laser is made incident on the fiber through a phase mask of a period as per the design of the FBG being inscribed and scans on the required length of the fiber to get the FBG inscription of the designed parameters e.g. (a typical value of 1060 nm periodicity phase mask is used in our lab to achieve the peak wavelength in the range of ~1550 nm). To achieve the FBG with more than 90% reflectivity, 7-8 scans were applied. For LPG fabrication point-by point method is used to expose fiber core to laser radiation. To monitor the FBG parameter within the appropriate design, optical spectrum analyzer with broadband ASE (Amplified Stimulated Emission) source is used while for LPGs white light source is employed. Figure shows the experimental setup for grating fabrication.

Grating fabrication set up.

To protect the FBG from external environment and to provide mechanical strength, the stripped fiber section is immediately re-coated with acrylate while LPGs are left uncoated so that they remain sensitive to external refractive index. Care should be taken to keep the bare LPGs well protected; they should be either fixed in a working glass cell or kept properly covered when not in use. Finally, for stabilization of grating properties over long period of time, thermal annealing of inscribed gratings at high temperature (>150 °C) is carried out.

Current Research Scenario

The multiplexing, multi-parameter and minimally invasive sensing capabilities even in high electric/magnetic field environments of fiber grating sensors make them befitting for various applications in healthcare as evident by past and ongoing research activities all over the world. Some areas of current activities are described briefly in subsequent sub-sections.

Biomechanics

Biomechanics involves application of engineering mechanics to biological systems to better understand them and make use of conventional electrical strain gages (ESGs) as a tool. These ESGs, considered gold standard for strain measurement, consist of metallic parts and it is difficult to make them adhere on the surface of a bone or other biological tissues. Besides, working of an ESG depends upon measuring electric resistance that varies with applied strain and so it is not suitable in strong electric and magnetic field environment associated with medical appliances. Moreover, they cannot be made completely biocompatible. Fiber gratings have an edge over the conventional electrical gages even for in-vivo applications because they have smaller risk for infection and can also be used even on curved surfaces or in locations where use of a conventional gage is technically and medically not feasible.

A temperature independent array of FBGs with proper design can be used as pressure sensor to measure muscular strength of hands or weight profile of patients when used under foot. In a study started almost a decade earlier, FBG pressure sensors embedded in Carbon Fiber reinforced material were designed and investigated by a research group in Singapore for monitoring the foot pressure of diabetic patients. These sensors give better results in terms of sensitivity, cost and compactness as compared to conventional foot pressure sensing systems. This concept along with development of multiple neural networks for continuous monitoring of various parameters simultaneously can be used for human gait analysis. Gait analysis is important for patients with cerebral palsy, neuromuscular disorders or diabetes and many hospitals worldwide now have gait labs both to design treatment plans, and for follow-up monitoring. Existing systems of gait analysis use large numbers of sensors or complex imaging systems, whereas several FBGs on a single strand of fiber with one interrogation unit can be embedded within materials forming a surface without loss of material strength. Recently, FBG sensors have been investigated for distributed tactile sensing in a grid arrangement along with a neural network to detect the position and shape of a contacting load simultaneously in real- time.

Multiple FBG sensors in a single fiber can be bonded at strategic locations on the patients' bed to continuously monitor their movements from a remote station. The concept of a smart bed is under investigation in Singapore. A series of 12 FBG sensors underneath the patient's mattress with suitable algorithm give pressure profile and respiratory rate of the patient while another set of gratings placed on top of the mattress detect heart rate. Existing methods require different techniques for each individual parameter while this single system shown schematically in figure, can monitor respiratory rate, heart rate, pressure points and occupancy of patient on bed in a continuous, nonintrusive and robust manner.

An ongoing project entitled, "Intelligent adaptable surface with optical fiber sensing for pressure- tension relief" (IASIS) of European Commission is incorporating FBG-based 2D and 3D sensing systems for utilization in "smart" fiber-based Human Machine Interfaces (HMI) employed in clinical beds, amputee sockets and wheelchair seating systems, targeting pressure ulcer and wound treatment.

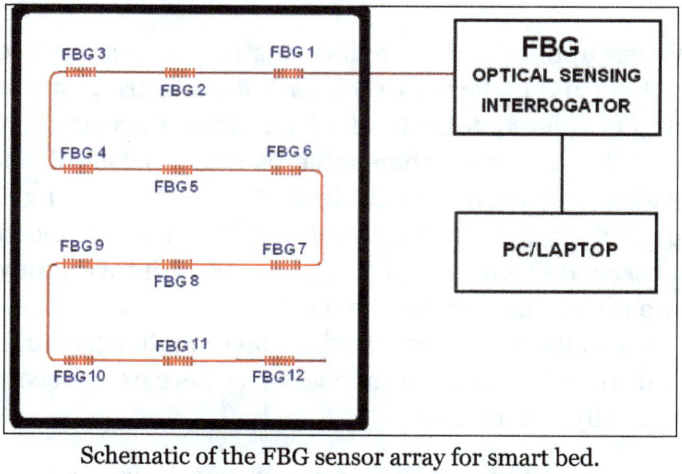

Schematic of the FBG sensor array for smart bed.

Recently some studies on the possible use of FBG as a strain gage in bones have been undertaken. Although in-vivo strain measurement in humans is not very common, the researchers at Hadassah University Hospital, Israel have reported use of instrumented bone staples made of electrical strain gages in some volunteers. Talaia et al have first reported use of FBG sensor array to study strains in fracture fixation of synthetic femur. As strain measurement on bone plates using ESG is technically difficult and not feasible, FBGs are good alternative to assess the stiffness of callus formation of fractured bones. Use of FBG sensors in place of ESGs have been validated to measure deformation in human cadaver femur bone specimen under in-vitro loading condition. The effect of decalcification on strain response of a goat tibia was investigated in-vitro using FBG sensors by our group. In the investigation, two tibia bone samples were taken from the same animal. The FBG sensor was directly bonded onto the surface at the midpoint of the bone shaft, which is the most vulnerable point, using standard cynoacrylate adhesive and cured properly. One sample was decalcified in gradual steps while the other was kept in saline solution for a comparative study. Both the bone samples were strained by using three-point bending technique and corresponding Bragg wavelength shifts were recorded. Strain response of the decalcified and untreated bones was studied concurrently to monitor the effects of calcium loss and that of degradation with time. The strain generated for the same stress increased with greater degree of decalcification e.g. calcium loss of even 0.3906 gm (treatment 1) resulted in 1.3 times/ 24% more strain for the same load while calcium loss of 1 gm resulted in 50% increase in strain and when calcium loss was 2.78 gm the increase in strain reached more than 2000% i.e. 22 times more strain as compared to that before decalcification. Figures show the schematic of the experimental setup developed and the results obtained respectively. The Dexa results of bone samples matched with our results for mineral loss.

It is possible to measure strain less than 5 micro-strains accurately using this sensing technique, which can be indicative of the onset of decalcification. The objective of this investigation was to develop a different, efficient and safe method to estimate calcium levels in bones. The small size of fiber can be utilized to make strain staples much smaller then the existing ones so that it can be implanted using minimally-invasive surgical method, or, as this technology is still developing it may advance into non-invasive method in future.

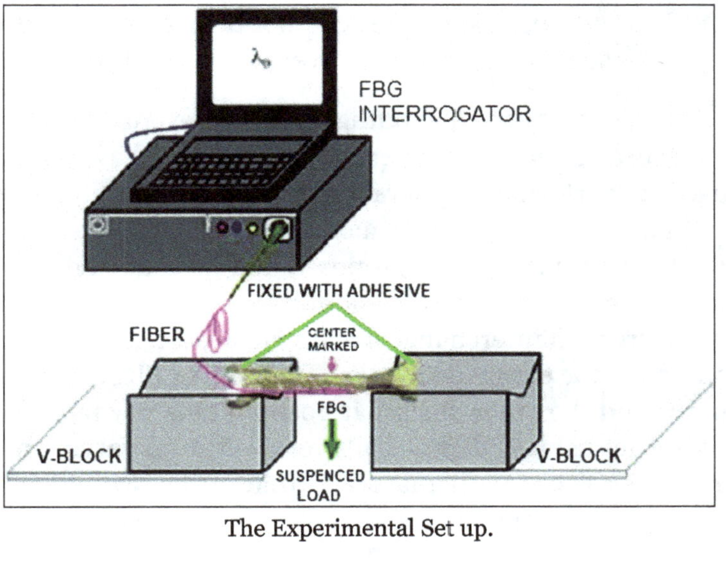

The Experimental Set up.

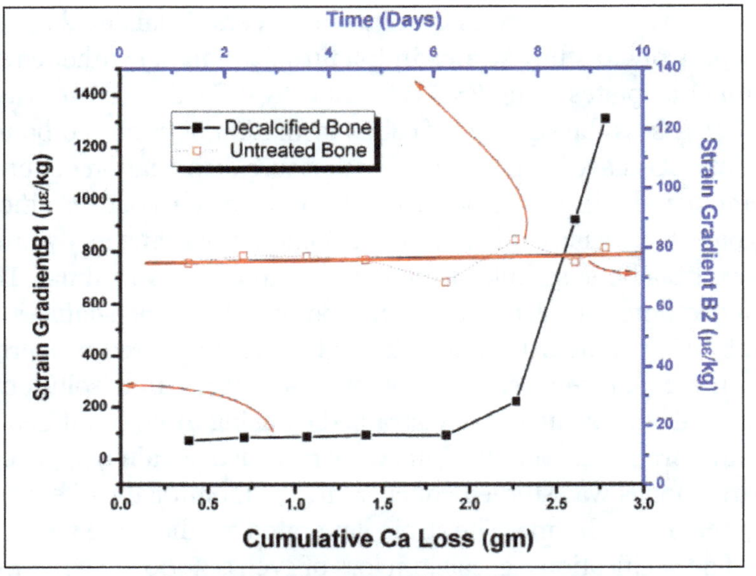

The Results of Bone Decalcification Study.

First application of embedded FBG sensors in dental biomechanics was reported to monitor the force and temperature as a function of time in dental splints used by patients with obstructive sleep apnoea. This monitoring gives a clear indication of the compliance of the patient with regard to the proper usage of splint which is necessary for its effectiveness. In another type of investigation FBGs written in polarization maintaining fibers were used to monitor the drying of dental cement and the corresponding stress build-up. To measure strain at a mandible surface caused by impact loads on dental implants, an FBG sensor was employed by JCC Silva et al on a dried cadaveric mandible showing the feasibility of using FBG to monitor dynamic strain in complex biomechanical systems. A research group in Portugal has applied FBG sensors to assess the performance of dental implant system by measuring static and dynamic bone strains around it. Conventional techniques can not be used to measure strains inside bones. This study can lead to significant improvement in the design of dental implants.

Now-a-days, the use of custom-made mouthguards as preventive measures for persons participating in sports activities is being encouraged as they have an injury-preventing ability. To evaluate the performance characteristics of such mouthguards, no standard technique exists till date. A unique experimental scheme utilizing fiber Bragg gratings (FBGs) as distributed strain sensors is proposed and investigated by our group to estimate impact absorption capability of custom-made mouthguards. In the scheme, a pendulum based fixture with interchangeable impact load e.g. cricket, hockey and steel balls, was custom made for the investigation. Two sets of FBGs were used; one at the mouthguard surface and the other at similar position on a jaw model. The fixture was used to simulate impact using different balls released from varying angles and Bragg wavelength shifts of FBGs at mouthguard and that at the jaw model were recorded. Figures show photograph of the jaw model, mouthguard with pendulum setup and change in FBG spectrum due to impact. It is obvious from the FBG response shown in figure that for the same impact there is no detectable change in the Bragg wavelength of the FBG bonded on the jaw model while FBG on the mouthguard has shown large wavelength shift with each impact. Relative Bragg wavelength shift with respect to each impact load determines the protection capability of the mouthguard. Due to multiplexing capability of FBGs, it is possible to fix multiple sensors in series at various points of mouthguard and denture to evaluate the effect of a single impact on different locations simultaneously. Impact tests on various locations can detect the vulnerable points where the mouthguard is less protective.

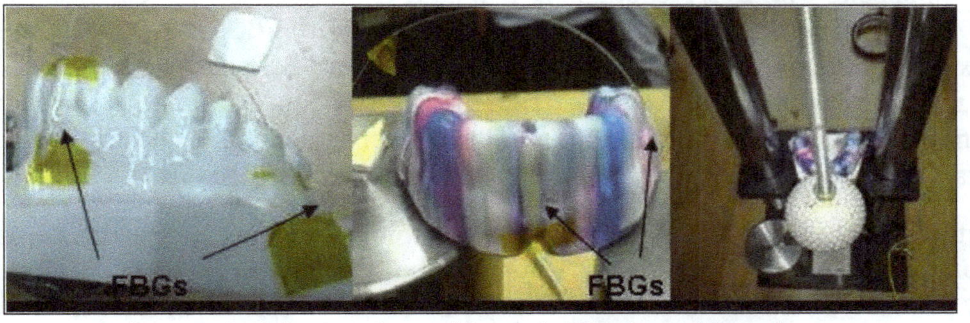

Photograph of the Experimental Set up for Mouthguard Experiment.

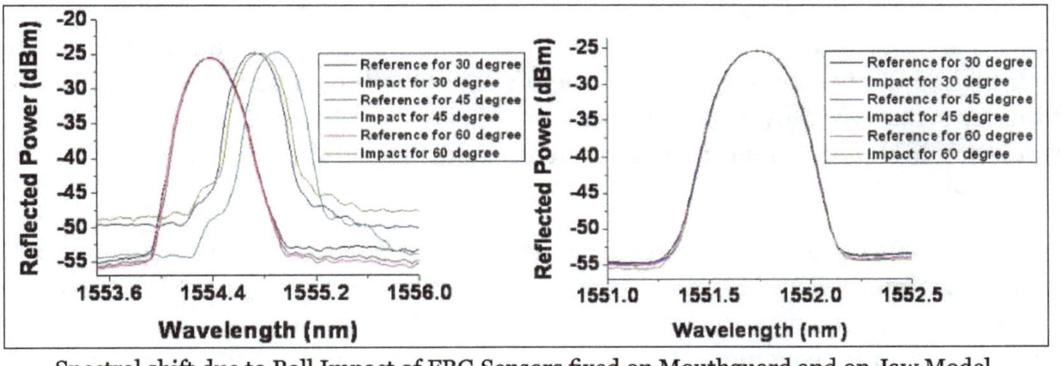

Spectral shift due to Ball Impact of FBG Sensors fixed on Mouthguard and on Jaw Model.

Through such studies it will be possible to quantify the level of protection and hence to predict the required modifications in the mouthguard. This research work thus, is important for the establishment of guidelines for design of safer mouthguards.

J Paul et al had suggested use of five FBGs written at different wavelengths to measure handgrip strength through a grip holder. Handgrip strength monitoring is rated as one of the top ten fitness tests to evaluate different physical and functional disorders related to healthcare. The conventional methods (dynamometer) are rough, uncomfortable and do not provide individual finger strengths; thus not suitable especially for rehabilitation programs.

Researchers have reported an FBG based sensor in instrumented tibial spacer (ITS) to correct misalignment during total knee replacement surgery. The sensor, comprising of optical fibers with sampled chirped gratings inscribed on each fiber to generate a pressure profile, was embedded in a fiberreinforced composite. During a total knee joint replacement procedure, the ITS sensor can slide in place of the prosthetic spacer. The femur can be rolled over the ITS sensor and the alignment checked from the pressure map displayed. Any misalignment can be corrected with repeated checking. After the measurements are taken and the required alignment achieved, the ITS sensor can be replaced by the actual tibial prosthetic spacer and the knee joint can be sutured.

Methods were developed by Dennison et al 2007 to measure intervertebral disc pressure response to compressive load in five lumbar functional spine units, using FBG in a patented configuration. The pressure measurement with FBGs is less disruptive than the existing techniques. In an improved configuration FBG sensor placed in silicone filled needle were applied to intervertebral disc pressure measurements in a cadaveric porcine functional spinal unit and the results were in agreement to those obtained with the standard strain gage sensor.

Investigative study of FBG sensor for in-vitro biomechanical properties of porcine tendons was reported by Miloslav Vilimek. The tendon force was calibrated using Bragg wavelength measurement of the FBG bonded on the tendon with applied load. FBG was used as displacement sensor on cadaver Achilles tendon and knee ligament for movement measurement of tendons and ligaments. Study of change in length of ligament under various strain conditions is important as ligaments experience much higher strain as compared to bone for same loading. The FBG sensors exhibited higher sensitivity, low noise and same accuracy as compared to stereo-optic measurement which are though non-invasive have limitations of poor accuracy and high noise level.

In a very recent development a research group from Portugal has investigated osteoblastic biocompatibility of optical fibers and stability of the properties of FBG sensors for their invivo usage. The study analysed the behaviour of human bone marrow cells cultured in osteogenic-induced conditions over an optical fiber and in parallel, the sensing capability of FBG sensors throughout the culture time was assessed. The results showed that in addition to the excellent osteoblastic cytocompatibility, FBGs maintained the physical integrity and functionality, as their sensing capability was not affected throughout the cell culture period. Results suggest the possibility of in-vivo osseointegration of the optical fiber/FBGs anticipating a variety of applications in bone mechanical dynamics.

Biosensors

The ability of LPGs to detect refractive index variation in their vicinity has great potential for detection of clinical analytes and can be made to detect extremely low concentrations. An LPG with an immobilized antibody film on its surface is a very efficient device to detect target antigen bonding to this film by means of refractive index change associated with the process. The advantage of using LPG is that it is a direct and label free sensor which does not require any additional reagents to visualize binding. Figure indicates the basic experimental set up for an LPG based biosensor. DeLisa et al have first reported use of LPG as biosensor for detection of human IgG by specific antibody-antigen binding with immobilized goat anti-human IgG antibody on the chemically treated surface of LPG. The system could work for antigen solution concentration between 2-100 ⊣ g mL^{-1}

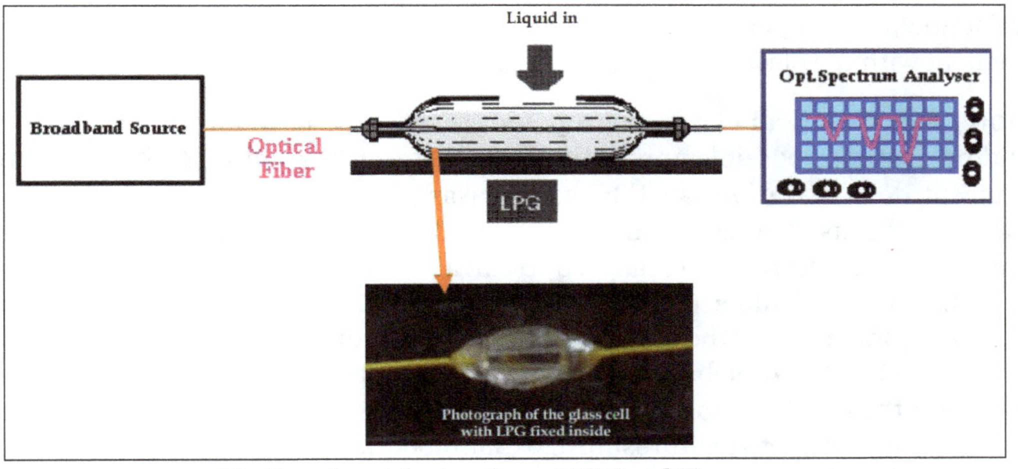

The Experimental set up for an LPG Based Biosensor.

Biosensor system though the product is yet to be commercialized. The sensitivities of these LPG sensors were found to be comparable to those of ELISA techniques. LPG sensors have also been used for monitoring microbial activity. Higher sensitivity in LPG sensors can be achieved by using gratings with smaller period or reduced- diameter cladding. Chen et al have verified high sensitivity of smaller period LPGs and their reusability by detecting DNA hybridization.

In normal FBGs as the optical signals are confined to the fiber core regions they are insensitive to external refractive index variation, but they can be made sensitive either by writing tilted grating or etching the cladding part. A Bovine Serum Albumin (BSA) immobilized tilted FBG based immunosensor was reported recently to detect anti-BSA. In another approach, highly sensitive etched FBG sensors have been demonstrated by Chryssis et al to detect hybridization of single strand DNA. All the FBG based biosensing research is currently at the level of laboratory investigations only as it requires either specially designed or etched FBGs that are difficult if not impractical to fabricate and expensive as compared to LPGs.

FBG/LPG sensor systems can be incorporated with the microchips to detect chemical changes in nanolitre amounts of liquids and have the potential of being part of a μ-TAS (micro total analysis system) or lab-on-chip. Miniaturization and integration of light sources, sensors, detectors, as well as the corresponding signal processing is required for implementation of these concepts in a practical analysis system.

Cardiology

Flow-directed thermodilution catheters with conventional thermistor and thermocouple devices were commercially available for many years to monitor heart efficiency. However, as these sensors are electrically active they are not appropriate for use in a number of medical applications, especially in high magnetic fields associated with NMR machines. Rao and coworkers in UK used a catheter with FBG temperature sensor in place of thermocouple in a test rig set up to simulate blood flow in the heart with a peristaltic pump to simulate the heart pump. The results were found to be in good agreement with the electrical sensor with smoother temperature profile.

In 2005, Deniz Gurkan et al had proposed FBG sensor to monitor heartbeat using sound pressure for possible application in ballistocardiography (BCG). For a proof-of-concept demonstration, an FBG sensor was glued to vibrating membrane of a subwoofer of a speaker set. Using recordings of various heartbeat sounds, spectral changes were monitored and analyzed to extract all relevant information. They predicted that in a real life scenario, the FBG would be placed on the patient's body in the same way as a stethoscope to detect any abnormalities in the heart muscle more efficiently. FBG for blood pressure monitoring had been reported by Brakel et al where they had proposed FBG FabryPerot interferometer (FBGI) as a sensor configuration to detect strain resulting from blood pressure applied to the walls of an artery situated near the patient's skin. In an investigation they demonstrated an optical blood pressure manometer that not only measured accurate systolic and diastolic blood pressures once it was calibrated, but also provided a continuous pressure waveform quite comparable with conventional Sphygmomanometer pulse wave velocity system readings.

FBG sensors, being small and light, have particular advantages for use in endoscopic applications and offer the possibility of being used in conjunction with MRI scanning, thus opening up an opportunity for performing surgery with continual scanning. FBGs can be mounted on the tip of an intra-aortic catheter that serves as a laser-ablation delivery probe for the treatment of atrial fibrillation. The FBGs can give real-time, objective measure of tip-to-tissue contact force during the catheter ablation procedure. Force control is essential for delivering appropriate laser ablation pulses needed to produce lesions that are induced in the heart walls. If the contact force is too great, the catheter tip may perforate the heart wall and if it is too light, the procedure may be ineffective. Such force-sensing catheters with accompanying system have undergone extensive pre-clinical and clinical validation in the United States & Europe and are currently undergoing precertification trials.

Gynecology

Measurement of the pelvic muscles pressure was demonstrated using FBG based intravaginal probe in Portugal. This measurement is essential for understanding pelvic floor disorders pathophysiology. The system was tested in a sample of patients with known pelvic floor disorders and the preliminary investigations indicated good sensitivity to radial pressure changes within the pelvic floor due to normal breathing cycle of the patient.

High Intensity Ultrasonic Field Measurements

High-intensity ultrasonic fields are used in various medical applications like ultrasound surgery, hyperthermia, lithotripsy and even diagnostic ultrasound. Thus safety concerns demand their accurate level assessment. The conventional detection techniques utilizing piezoelectric devices are susceptible to electromagnetic interference and signal distortion. Fisher et al demonstrated that FBG sensors can be used for this purpose by implementing them to detect signals of frequencies in the range of 500 kHz and 4 MHz.

Respiration Monitoring System

A temperature independent FBG or LPG can be designed for non-invasive measurement of the torso movement during respiration ventilatory movements to understand respiratory physiology and to monitor the lung function. In a preliminary work, Günther Wehrle et al in 2001 employed FBG sensor on the chest using an elastic belt to hold it in place to detect thorax movements during artificial ventilation, even in the presence of electrical bursts caused by electrodes situated on the chest. Expansion of the thorax cage during respiration was accordingly transmitted to the sensor grating and caused it to deform under the strain. First application of a multiplexed LPG array on curvature sensing garment used to monitor the thoracic and abdominal movements of a human during respiration was reported by T. Allsop. They have shown that, it is possible to generate a geometric profile of the chest and abdomen in three dimensions with an array of 20 sensors.

Temperature Monitoring

Thermal therapy involves destruction of redundant tissue by heating or freezing without surgery. Examples of thermal therapies include treatment of benign prostatic diseases, ablation therapy of cardiac arrhythmia, microwave induced hyperthermia for radio therapy in the cancer treatment, during which temperature of both diseased and normal tissues should ideally be monitored. Most temperature sensors measure temperature at one point only, whereas FBG sensors because of their multiplexing capability provide temperature distribution during thermal treatment using one probe only. Temperature monitoring of a patient during MRI procedure is crucial in general and in particular to ensure that the implantable devices do not heat up under strong fields of MRI. An FBG probe can be placed inside a NMR machine with high magnetic field for remote temperature measurement.

An FBG based temperature profile monitoring system was first proposed by researchers from UK and Australia using four strain-free FBG array that meets the established medical requirements and tested the feasibility of FBG temperature probe inside a NMR machine with high magnetic field and established its working. The first in-vivo trials of such temperature probes incorporating five FBGs along a single fiber were undertaken successfully at the Cancer Research Institute in Perth, Australia by the same group on rabbits undergoing hyperthermia treatment of the kidney and liver via inductive heating of metallic implants. A distributed FBG sensor system was tested effectively in temperature range of -195.8 °C to100 °C for in-vivo use during freezing of porcine liver, for their mechanical stability and MR compatibility.

FBGs are very useful in cryosurgery for localized cancer treatment, where precision of millimeters matters. It is very critical to monitor exact temperature profile throughout surgical procedure to ensure adequate freezing (-40°C) in the cancerous region but preventing low temperatures in the attached nerve fibers, blood vessels or nearby organs for their safety. An FBG based re-usable multi-point temperature monitoring system was developed recently which can record temperatures in a continuum of either four points at 10-mm intervals, or eight points at 5-mm intervals.

FBG based Manometry Catheter

A research group from Australia has demonstrated multi-element catheter using FBG pressure sensor in in-vivo trials in the human oesophagus for diagnosis of gastrointestinal motility disorders. The FBG based catheters had an outer diameter of less than 3 mm and were very flexible in comparison to solid state catheters, allowing easy nasal intubation and relative comfort for the patient during subsequent clinical diagnosis. Also, due to multiplexing capability of FBGs a large number of sensors can be accommodated in a single fiber thus making it possible to extend the total span of the sensing region, with no change in the catheter size. For validation, 10 healthy volunteers were intubated via the nose with both catheter assemblies and asked to perform series of controlled water swallows. The results were directly compared with a commercially available solid state catheter and largely equivalent response was obtained.

Smart Clothing

Smart clothings have the capability to sense stimuli from the environment, to react and adapt to them. Due to their small size and flexibility fiber optic sensors can be woven into the textile and be an integral part of the smart system. This type of wearable and smart textile is useful not only for patient health monitoring but also for soldiers in battlefields, people in sports and even for general safety of public. In a collaborative work 10 expert groups from 5 EU countries have initiated a novel project, i.e. Optical Fiber Sensors Embedded into Technical Textile for Healthcare (OFSETH) to create smart wearable textiles with FBG along with other optical devices for measurements of various vital parameters 6. Though preliminary experiments with the textile containing FBGs had been undertaken further investigations are needed for practical implementation of smart textile concept in terms of better integration of the fiber with textile, of different patient profiles and overall performance in real environment.

Robotic Surgery

A force sensing robot fingers using embedded FBG sensors and shape deposition manufacturing have been demonstrated recently. With their very small size and highly precise real-time measurements FBG sensors have the potential to be a part of surgical robot. Shape-sensing systems can be created by using arrays of FBGs distributed along multicore, singlemode fibers that will help determine the precise position and shape of medical tools and robotic arms used during minimally invasive/robotic surgery.

Optical Coherence Tomography (OCT)

OCT is an upcoming technology for non-invasive cross-sectional imaging in biological systems with a great potential for morphological assessment of various tissues. OCT acquires very high-resolution images through the detection of backscattered near infrared light with the potential to identify a wide range of pathologies in-vivo. Clinical applications of OCT are constrained by its limited penetration depth in tissue and miniaturized fiber-optic probes are used to image deeper within the body.

FBGs with precise Bragg wavelength, narrow spectral bandwidth less than 0.1 nm and high reflectivity are being applied in the OCT systems to calibrate and align the custom-made spectrometer for the spectral-domain optical coherence tomography (SD-OCT) system operating in the 1300-nm wavelength range. The calibration and characterization protocol was presented and investigated very recently.

2

Biomedical Engineering

Biomedical engineering is an interdisciplinary field of engineering and medicine that applies engineering design and principles to medicine and biology for diagnostic and therapeutic purposes. Biomimetics, neural engineering, bioinformatics, biomechanics, etc. fall under its domain. This is an introductory chapter which will briefly introduce about biomedical engineering.

Biomedical engineering is a discipline that advances knowledge in engineering, biology and medicine, and improves human health through cross-disciplinary activities that integrate the engineering sciences with the biomedical sciences and clinical practice. It includes:

- The acquisition of new knowledge and understanding of living systems through the innovative and substantive application of experimental and analytical techniques based on the engineering sciences.

- The development of new devices, algorithms, processes and systems that advance biology and medicine and improve medical practice and health care delivery.

The term "biomedical engineering research" is thus defined in a broad sense: It includes not only the relevant applications of engineering to medicine but also to the basic life sciences.

Development of Bioengineering

Over the last few years there has been a major paradigm shift in both Europe and the United States away from traditional schemes of health care towards health care systems which are much more dependent on technology. This is true in terms of diagnosis (eg body scanners); treatment (radiation therapy and minimal access surgery); and health care system integration (via information technology).

In parallel with these changes, there has been a progressive increase in the proportion of the national Gross Domestic Product spent in the medical sector. For example, in the United Kingdom it is currently between 6 and 7%, in Germany about 9%, and in the United States about 14%. This has resulted partly from demographic changes and additionally from increasing public demand for better health care.

As medical practice becomes more technologically based, a progressive shift is occurring in industry to meet the demand. Developments in science and engineering are increasingly being directed away from traditional technologies towards those required for health care in its widest sense. Although in many countries there is a problem with escalating costs in the medical sector, technology can contribute to economies because of falling costs of electronic/physics based components relative to those for personnel, and because of technologically based screening programmes.

Speciality Areas

Some of the well established Speciality areas within the field of biomedical engineering are bioinstrumentation, biomechanics, biomaterials, systems physiology, clinical engineering, and rehabilitation engineering.

Bioinstrumentation is the application of electronics and measurement principles and techniques to develop devices used in diagnosis and treatment of disease. Computers are becoming increasingly important in bioinstrumentation, from the microprocessor used to do a variety of small tasks in a single purpose instrument to the extensive computing power needed to process the large amount of information in a medical imaging system.

Biomechanics is mechanics applied to biological or medical problems. It includes the study of motion, of material deformation, of flow within the body and in devices, and transport of chemical constituents across biological and synthetic media and membranes. Efforts in biomechanics have developed the artificial heart and replacement heart valves, the artificial kidney, the artificial hip, as well as built a better understanding of the function of organs and musculoskeletal systems.

Biomaterials describes both living tissue and materials used for implantation. Understanding the properties of the living material is vital in the design of implant materials. The selection of an appropriate material to place in the human body may be one of the most difficult tasks faced by the biomedical engineer. Certain metal alloys, ceramics, polymers, and composites have been used as implantable materials. Biomaterials must be nontoxic, noncarcinogenic, chemically inert, stable, and mechanically strong enough to withstand the repeated forces of a lifetime.

Systems physiology is the term used to describe that aspect of biomedical engineering in which engineering strategies, techniques and tools are used to gain a comprehensive and integrated understanding of the function of living organisms ranging from bacteria to humans. Modeling is used in the analysis of experimental data and in formulating mathematical descriptions of physiological events. In research, models are used in designing new experiments to refine our knowledge. Living systems have highly regulated feedback control systems which can be examined in this way. Examples are the biochemistry of metabolism and the control of limb movements.

Clinical engineering is the application of technology for health care in hospitals. The clinical engineer is a member of the health care team along with physicians, nurses and other hospital staff. Clinical engineers are responsible for developing and maintaining computer databases of medical instrumentation and equipment records and for the purchase and use of sophisticated medical instruments. They may also work with physicians on projects to adapt instrumentation to the specific needs of the physician and the hospital. This often involves the interface of instruments with computer systems and customized software for instrument control and data analysis. Clinical engineers feel the excitement of applying the latest technology to health care.

Rehabilitation engineering is a new and growing Speciality area of biomedical engineering. Rehabilitation engineers expand capabilities and improve the quality of life for individuals with physical impairments. Because the products of their labor are so personal, often developed for particular individuals or small groups, the rehabilitation engineer often works directly with the disabled individual.

These Speciality areas frequently depend on each other. Often the biomedical engineer who works in an applied field will use knowledge gathered by biomedical engineers working in more basic areas. For example, the design of an artificial hip is greatly aided by a biomechanical study of the hip. The forces which are applied to the hip can be considered in the design and material selection for the prosthesis. Similarly, the design of systems to electrically stimulate paralyzed muscle to move in a controlled way uses knowledge of the behavior of the human musculoskeletal system. The selection of appropriate materials used in these devices falls within the realm of the biomaterials engineer. These are examples of the interactions among the Speciality areas of biomedical engineering.

TISSUE ENGINEERING

Tissue engineering is the use of a combination of cells, engineering, and materials methods, and suitable biochemical and physicochemical factors to improve or replace biological tissues. Tissue engineering involves the use of a tissue scaffold for the formation of new viable tissue for a medical purpose. While it was once categorized as a sub-field of biomaterials, having grown in scope and importance it can be considered as a field in its own.

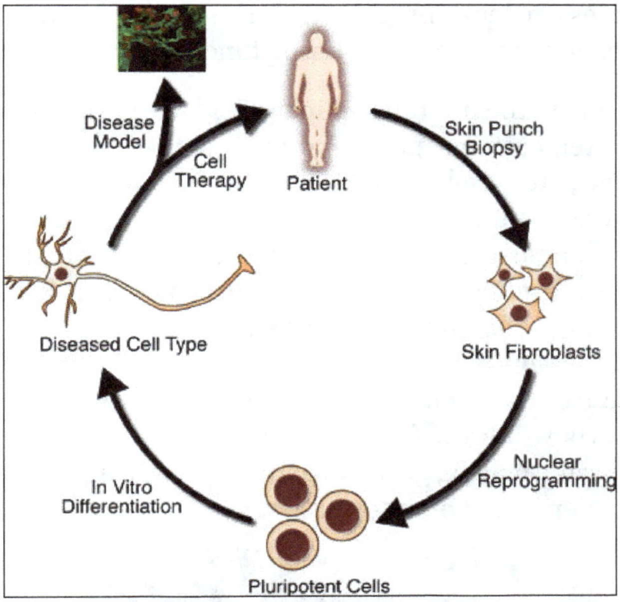

A simplified overview of the general methods used in regenerative medicine.

While most definitions of tissue engineering cover a broad range of applications, in practice the term is closely associated with applications that repair or replace portions of or whole tissues (i.e., bone, cartilage, blood vessels, bladder, skin, muscle etc.). Often, the tissues involved require certain mechanical and structural properties for proper functioning. The term has also been applied to efforts to perform specific biochemical functions using cells within an artificially-created support system (e.g. an artificial pancreas, or a bio artificial liver). The term *regenerative medicine* is often used synonymously with tissue engineering, although those involved in regenerative medicine place more emphasis on the use of stem cells or progenitor cells to produce tissues.

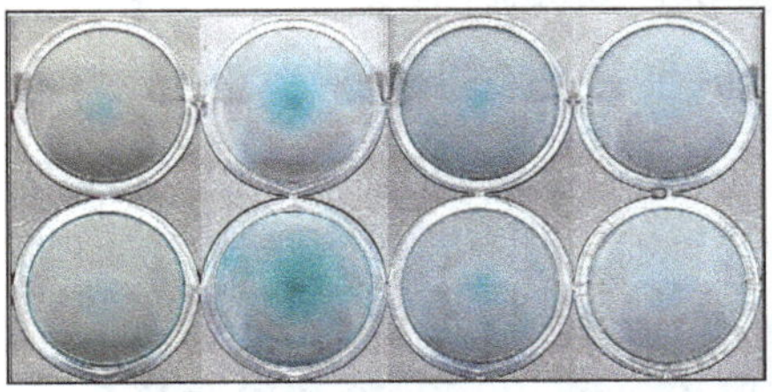

Micro-mass cultures of C3H-10T1/2 cells at varied oxygen tensions stained with Alcian blue.

A commonly applied definition of tissue engineering, as stated by Langer and Vacanti, is "an interdisciplinary field that applies the principles of engineering and life sciences toward the development of biological substitutes that restore, maintain, or improve function or a whole organ". Tissue engineering has also been defined as "understanding the principles of tissue growth, and applying this to produce functional replacement tissue for clinical use". A further description goes on to say that an "underlying supposition of tissue engineering is that the employment of natural biology of the system will allow for greater success in developing therapeutic strategies aimed at the replacement, repair, maintenance, or enhancement of tissue function".

Powerful developments in the multidisciplinary field of tissue engineering have yielded a novel set of tissue replacement parts and implementation strategies. Scientific advances in biomaterials, stem cells, growth and differentiation factors, and biomimetic environments have created unique opportunities to fabricate tissues in the laboratory from combinations of engineered extracellular matrices ("scaffolds"), cells, and biologically active molecules. Among the major challenges now facing tissue engineering is the need for more complex functionality, as well as both functional and biomechanical stability and vascularization in laboratory-grown tissues destined for transplantation. The continued success of tissue engineering and the eventual development of true human replacement parts will grow from the convergence of engineering and basic research advances in tissue, matrix, growth factor, stem cell, and developmental biology, as well as materials science and bioinformatics.

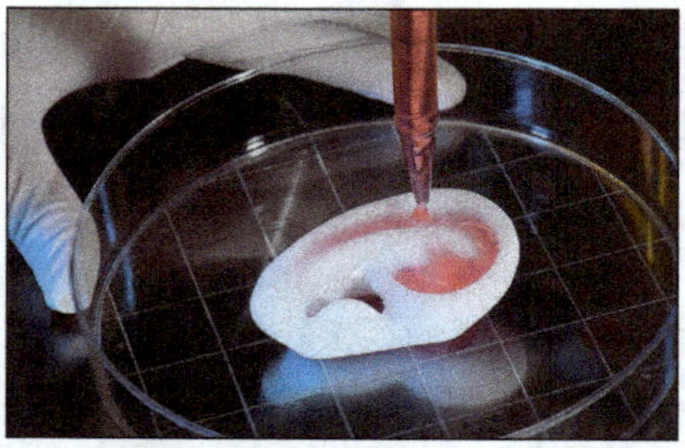

Regenerating a human ear using a scaffold.

Examples:

- In vitro meat: Edible artificial animal muscle tissue cultured *in vitro*.
- Bioartificial liver device: Several research efforts have produced hepatic assist devices utilizing living hepatocytes.
- Artificial pancreas: Research involves using islet cells to produce and regulate insulin, particularly in cases of diabetes.
- Artificial bladders: Anthony Atala (Wake Forest University) has successfully implanted artificially grown bladders into seven out of approximately 20 human test subjects as part of a long-term experiment.
- Cartilage: Lab-grown tissue was successfully used to repair knee cartilage.
- Scaffold-free cartilage: Cartilage generated without the use of exogenous scaffold material. In this methodology, all material in the construct is cellular or material produced directly by the cells themselves.
- Doris Taylor's heart in a jar.
- Tissue-engineered airway.

- Tissue-engineered vessels.
- Artificial skin constructed from human skin cells embedded in a hydrogel, such as in the case of bioprinted constructs for battlefield burn repairs.
- Artificial bone marrow.
- Artificial bone.
- Laboratory-grown penis.
- Oral mucosa tissue engineering.
- Foreskin.

Cells as Building Blocks

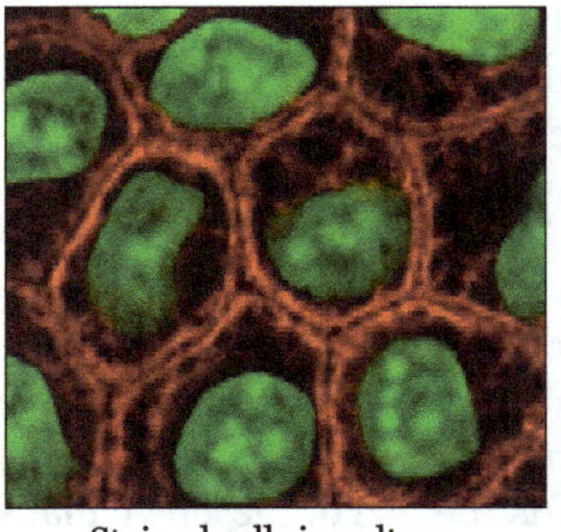

Stained cells in culture.

Tissue engineering utilizes living cells as engineering materials. Examples include using living fibroblasts in skin replacement or repair, cartilage repaired with living chondrocytes, or other types of cells used in other ways.

Cells became available as engineering materials when scientists at Geron Corp. discovered how to extend telomeres in 1998, producing immortalized cell lines. Before this, laboratory cultures of healthy, noncancerous mammalian cells would only divide a fixed number of times, up to the Hayflick limit, before dying.

Extraction

From fluid tissues such as blood, cells are extracted by bulk methods, usually centrifugation or apheresis. From solid tissues, extraction is more difficult. Usually, the tissue is minced and then digested with the enzymes trypsin or collagenase to remove the extracellular matrix (ECM) that holds the cells. After that, the cells are free floating, and extracted using centrifugation or apheresis.Digestion with trypsin is very dependent on temperature. Higher temperatures digest the matrix faster but create more damage. Collagenase is less temperature dependent, and damages fewer cells, but takes longer and is a more expensive reagent.

Types of Cells

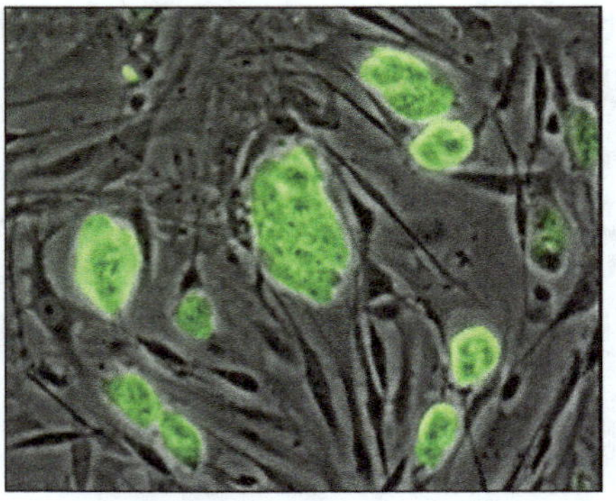

Mouse embryonic stem cells.

Cells are often categorized by their source: *Autologous* cells are obtained from the same individual to which they will be reimplanted. Autologous cells have the fewest problems with rejection and pathogen transmission, however, in some cases might not be available. For example, in genetic disease suitable autologous cells are not available. Also, very ill or elderly persons, as well as patients suffering from severe burns, may not have sufficient quantities of autologous cells to establish useful cell lines. Moreover, since this category of cells needs to be harvested from the patient, there are also some concerns related to the necessity of performing such surgical operations that might lead to donor site infection or chronic pain. Autologous cells also must be cultured from samples before they can be used: this takes time, so autologous solutions may not be very quick. Recently there has been a trend towards the use of mesenchymal stem cells from bone marrow and fat. These cells can differentiate into a variety of tissue types, including bone, cartilage, fat, and nerve. A large number of cells can be easily and quickly isolated from fat, thus opening the potential for large numbers of cells to be quickly and easily obtained.

Allogeneic cells come from the body of a donor of the same species. While there are some ethical constraints to the use of human cells for *in vitro* studies, the employment of dermal fibroblasts from human foreskin has been demonstrated to be immunologically safe and thus a viable choice for tissue engineering of skin.

Xenogenic cells are these isolated from individuals of another species. In particular animal cells have been used quite extensively in experiments aimed at the construction of cardiovascular implants.

Syngenic or *isogenic* cells are isolated from genetically identical organisms, such as twins, clones, or highly inbred research animal models. *Primary* cells are from an organism. *Secondary* cells are from a cell bank.

Stem cells are undifferentiated cells with the ability to divide in culture and give rise to different forms of specialized cells. According to their source stem cells are divided into "adult" and "embryonic" stem cells, the first class being multipotent and the latter mostly pluripotent; some cells are totipotent, in the earliest stages of the embryo.

While there is still a large ethical debate related with the use of embryonic stem cells, it is thought that another alternative source - induced stem cells may be useful for the repair of diseased or damaged tissues, or may be used to grow new organs.

Scaffolds

Scaffolds are materials that have been engineered to cause desirable cellular interactions to contribute to the formation of new functional tissues for medical purposes. Cells are often 'seeded' into these structures capable of supporting three-dimensional tissue formation. Scaffolds mimic the extracellular matrix of the native tissue, recapitulating the *in vivo* milieu and allowing cells to influence their own microenvironments. They usually serve at least one of the following purposes: allow cell attachment and migration, deliver and retain cells and biochemical factors, enable diffusion of vital cell nutrients and expressed products, exert certain mechanical and biological influences to modify the behaviour of the cell phase. In 2009, an interdisciplinary team led by the thoracic surgeon Thorsten Walles implanted the first bioartificial transplant that provides an innate vascular network for post-transplant graft supply successfully into a patient awaiting tracheal reconstruction.

To achieve the goal of tissue reconstruction, scaffolds must meet some specific requirements. High porosity and adequate pore size are necessary to facilitate cell seeding and diffusion throughout the whole structure of both cells and nutrients. Biodegradability is often an essential factor since scaffolds should preferably be absorbed by the surrounding tissues without the necessity of surgical removal. The rate at which degradation occurs has to coincide as much as possible with the rate of tissue formation: This means that while cells are fabricating their own natural matrix structure around themselves, the scaffold is able to provide structural integrity within the body and eventually it will break down leaving the newly formed tissue which will take over the mechanical load. Injectability is also important for clinical uses. Recent research on organ printing is showing how crucial a good control of the 3D environment is to ensure reproducibility of experiments and offer better results.

Materials

Many different materials (natural and synthetic, biodegradable and permanent) have been investigated. Most of these materials have been known in the medical field before the advent of tissue engineering as a research topic, being already employed as bioresorbable sutures. Examples of these materials are collagen and some polyesters.

New biomaterials have been engineered to have ideal properties and functional customization: injectability, synthetic manufacture, biocompatibility, non-immunogenicity, transparency, nano-scale fibers, low concentration, resorption rates, etc. PuraMatrix, originating from the MIT labs of Zhang, Rich, Grodzinsky, and Langer is one of these new biomimetic scaffold families which has now been commercialized and is impacting clinical tissue engineering.

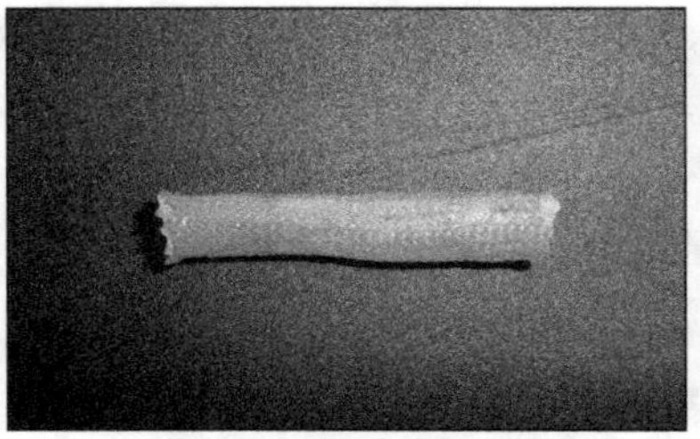

Tissue engineered vascular graft.

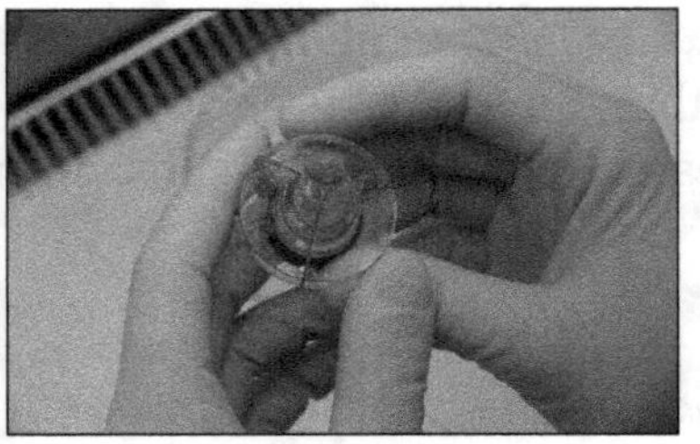

Tissue engineered heart valve.

A commonly used synthetic material is PLA - polylactic acid. This is a polyester which degrades within the human body to form lactic acid, a naturally occurring chemical which is easily removed from the body. Similar materials are polyglycolic acid (PGA) and polycaprolactone (PCL): Their degradation mechanism is similar to that of PLA, but they exhibit respectively a faster and a slower rate of degradation compared to PLA. While these materials have well maintained mechanical strength and structural integrity, they exhibit a hydrophobic nature. This hydrophobicity inhibits their biocompatibility, which makes them less effective for in vivo use as tissue scaffolding. In order to fix the lack of biocompatibility, much research has been done to combine these hydrophobic materials with hydrophilic and more biocompatible hydrogels. While these hydrogels have a superior biocompatibility, they lack the structural integrity of PLA, PCL, and PGA. By combining the two different types of materials, researchers are trying to create a synergistic relationship that produces a more biocompatible tissue scaffolding. Scaffolds may also be constructed from natural materials: in particular different derivatives of the extracellular matrix have been studied to evaluate their ability to support cell growth. Proteic materials, such as collagen or fibrin, and polysaccharidic materials, like chitosan or glycosaminoglycans (GAGs), have all proved suitable in terms of cell compatibility, but some issues with potential immunogenicity still remains. Among GAGs hyaluronic acid, possibly in combination with cross linking agents (e.g. glutaraldehyde, water-soluble carbodiimide, etc.), is one of the possible choices as scaffold material. Functionalized groups of scaffolds may be useful in the delivery of small molecules (drugs) to specific tissues. Another form of scaffold under investigation is decellularised tissue extracts whereby the remaining cellular remnants/extracellular matrices act as the scaffold. Recently a range of nanocomposites biomaterials are fabricated by incorporating nanomaterials within the polymeric matrix to engineer bioactive scaffolds.

Nanofiber Self-assembly

Molecular self-assembly is one of the few methods for creating biomaterials with properties similar in scale and chemistry to that of the natural *in vivo* extracellular matrix (ECM), a crucial step toward tissue engineering of complex tissues. Moreover, these hydrogel scaffolds have shown superiority in in vivo toxicology and biocompatibility compared to traditional macroscaffolds and animal-derived materials.

Textile Technologies

These techniques include all the approaches that have been successfully employed for the preparation of non-woven meshes of different polymers. In particular, non-woven polyglycolide structures have been tested for tissue engineering applications: Such fibrous structures have been found useful to grow different types of cells. The principal drawbacks are related to the difficulties in obtaining high porosity and regular pore size.

Solvent Casting and Particulate Leaching

Solvent casting and particulate leaching (SCPL) allows for the preparation of structures with regular porosity, but with limited thickness. First, the polymer is dissolved into a suitable organic solvent (e.g. polylactic acid could be dissolved into dichloromethane), then the solution is cast into a mold filled with porogen particles. Such porogen can be an inorganic salt like sodium chloride, crystals of saccharose, gelatin spheres or paraffin spheres. The size of the porogen particles will affect the size of the scaffold pores, while the polymer to porogen ratio is directly correlated to the amount of porosity of the final structure. After the polymer solution has been cast the solvent is allowed to fully evaporate, then the composite structure in the mold is immersed in a bath of a liquid suitable for dissolving the porogen: Water in the case of sodium chloride, saccharose and gelatin or an aliphatic solvent like hexane for use with paraffin. Once the porogen has been fully dissolved, a porous structure is obtained. Other than the small thickness range that can be obtained, another drawback of SCPL lies in its use of organic solvents which must be fully removed to avoid any possible damage to the cells seeded on the scaffold.

Gas Foaming

To overcome the need to use organic solvents and solid porogens, a technique using gas as a porogen has been developed. First, disc-shaped structures made of the desired polymer are prepared by means of compression molding using a heated mold. The discs are then placed in a chamber where they are exposed to high pressure CO_2 for several days. The pressure inside the chamber is gradually restored to atmospheric levels. During this procedure the pores are formed by the carbon dioxide molecules that abandon the polymer, resulting in a sponge-like structure. The main problems resulting from such a technique are caused by the excessive heat used during compression molding (which prohibits the incorporation of any temperature labile material into the polymer matrix) and by the fact that the pores do not form an interconnected structure.

Emulsification Freeze-drying

This technique does not require the use of a solid porogen like SCPL. First, a synthetic polymer is dissolved into a suitable solvent (e.g. polylactic acid in dichloromethane) then water is added to the polymeric solution and the two liquids are mixed in order to obtain an emulsion. Before the two phases can separate, the emulsion is cast into a mold and quickly frozen by means of immersion into liquid nitrogen. The frozen emulsion is subsequently freeze-dried to remove the dispersed water and the solvent, thus leaving a solidified, porous polymeric structure. While emulsification and freeze-drying allow for a faster preparation when compared to SCPL (since it does not require a time-consuming leaching step), it still requires the use of solvents. Moreover, pore size is relatively small and porosity is often irregular. Freeze-drying by itself is also a commonly employed technique for the fabrication of scaffolds. In particular, it is used to prepare collagen sponges: collagen is dissolved into acidic solutions of acetic acid or hydrochloric acid that are cast into a mold, frozen with liquid nitrogen and then lyophilized.

Thermally Induced Phase Separation

Similar to the previous technique, the TIPS phase separation procedure requires the use of a solvent with a low melting point that is easy to sublime. For example, dioxane could be used to dissolve polylactic acid, then phase separation is induced through the addition of a small quantity of water: a polymer-rich and a polymer-poor phase are formed. Following cooling below the solvent melting point and some days of vacuum-drying to sublime the solvent, a porous scaffold is obtained. Liquid-liquid phase separation presents the same drawbacks of emulsification/freeze-drying.

Electrospinning

Electrospinning is a highly versatile technique that can be used to produce continuous fibers from submicrometer to nanometer diameters. In a typical electrospinning set-up, a solution is fed through a spinneret and a high voltage is applied to the tip. The buildup of electrostatic repulsion within the charged solution, causes it to eject a thin fibrous stream. A mounted collector plate or rod with an opposite or grounded charge draws in the continuous fibers, which arrive to form a highly porous network. The primary advantages of this technique are its simplicity and ease of variation. At a laboratory level, a typical electrospinning set-up only requires a high voltage power supply (up to 30 kV), a syringe, a flat tip needle, and a conducting collector. For these reasons, electrospinning has become a common method of scaffold manufacture in many labs. By modifying variables such as the distance to collector, magnitude of applied voltage, or solution flow rate—researchers can dramatically change the overall scaffold architecture.

Historically, research on electrospun fibrous scaffolds dates back to at least the late 1980s when Simon showed that electrospinning could be used to produced nano- and submicron-scale fibrous scaffolds from polymer solutions specifically intended for use as *in vitro* cell and tissue substrates. This early use of electrospun lattices for cell culture and tissue engineering showed that various cell types would adhere to and proliferate upon polycarbonate fibers. It was noted that as opposed to the flattened morphology typically seen in 2D culture, cells grown on the electrospun fibers exhibited a more rounded 3-dimensional morphology generally observed of tissues *in vivo*.

CAD/CAM Technologies

Because most of the above techniques are limited when it comes to the control of porosity and pore size, computer assisted design and manufacturing techniques have been introduced to tissue engineering. First, a three-dimensional structure is designed using CAD software. The porosity can be tailored using algorithms within the software. The scaffold is then realized by using ink-jet printing of polymer powders or through Fused Deposition Modeling of a polymer melt.

A 2011 study by El-Ayoubi et al. investigated "3D-plotting technique to produce (biocompatible and biodegradable) poly-L-Lactide macroporous scaffolds with two different pore sizes" via solid free-form fabrication (SSF) with computer-aided-design (CAD), to explore therapeutic articular cartilage replacement as an "alternative to conventional tissue repair". The study found the smaller the pore size paired with mechanical stress in a bioreactor (to induce in vivo-like conditions), the higher the cell viability in potential therapeutic functionality via decreasing recovery time and increasing transplant effectiveness.

Laser-assisted Bioprinting

In a 2012 study, Koch et al. focused on whether Laser-assisted BioPrinting (LaBP) can be used to build multicellular 3D patterns in natural matrix, and whether the generated constructs are functioning and forming tissue. LaBP arranges small volumes of living cell suspensions in set high-resolution patterns. The investigation was successful, the researchers foresee that "generated tissue constructs might be used for in vivo testing by implanting them into animal models". As of this study, only human skin tissue has been synthesized, though researchers project that "by integrating further cell types (e.g. melanocytes, Schwann cells, hair follicle cells) into the printed cell construct, the behavior of these cells in a 3D in vitro microenvironment similar to their natural one can be analyzed", which is useful for drug discovery and toxicology studies.

Assembly Methods

One of the continuing, persistent problems with tissue engineering is mass transport limitations. Engineered tissues generally lack an initial blood supply, thus making it difficult for any implanted cells to obtain sufficient oxygen and nutrients to survive, or function properly.

Self-assembly

Self-assembly methods have been shown to be promising methods for tissue engineering. Self-assembly methods have the advantage of allowing tissues to develop their own extracellular matrix, resulting in tissue that better recapitulates biochemical and biomechanical properties of native tissue. Self-assembling engineered articular cartilage was introduced by Jerry Hu and Kyriacos A. Athanasiou in 2006 and applications of the process have resulted in engineered cartilage approaching the strength of native tissue. Self-assembly is a prime technology to get cells grown in a lab to assemble into three-dimensional shapes. To break down tissues into cells, researchers first have to dissolve the extracellular matrix that normally binds them together. Once cells are isolated, they must form the complex structures that make up our natural tissues.

Liquid-based Template Assembly

The air-liquid surface established by Faraday waves is explored as a template to assemble biological entities for bottom-up tissue engineering. This liquid-based template can be dynamically reconfigured in a few seconds, and the assembly on the template can be achieved in a scalable and parallel manner. Assembly of microscale hydrogels, cells, neuron-seeded micro-carrier beads, cell spheroids into various symmetrical and periodic structures was demonstrated with good cell viability. Formation of 3D neural network was achieved after 14-day tissue culture.

Additive Manufacturing

It might be possible to print organs, or possibly entire organisms using additive manufacturing techniques. A recent innovative method of construction uses an ink-jet mechanism to print precise layers of cells in a matrix of thermoreversible gel. Endothelial cells, the cells that line blood vessels, have been printed in a set of stacked rings. When incubated, these fused into a tube.

The field of three-dimensional and highly accurate models of biological systems is pioneered by multiple projects and technologies including a rapid method for creating tissues and even whole organs involve a 3D printer that can print the scaffolding and cells layer by layer into a working tissue sample or organ. The device is presented in a TED talk by Dr. Anthony Atala, M.D. the Director of the Wake Forest Institute for Regenerative Medicine, and the W.H. Boyce Professor and Chair of the Department of Urology at Wake Forest University, in which a kidney is printed on stage during the seminar and then presented to the crowd. It is anticipated that this technology will enable the production of livers in the future for transplantation and theoretically for toxicology and other biological studies as well.

Recently Multi-Photon Processing (MPP) was employed for in vivo experiments by engineering artificial cartilage constructs. An ex vivo histological examination showed that certain pore geometry and the pre-growing of chondrocytes (Cho) prior to implantation significantly improves the performance of the created 3D scaffolds. The achieved biocompatibility was comparable to the commercially available collagen membranes. The successful outcome of this study supports the idea that hexagonal-pore-shaped hybrid organic-inorganic microstructured scaffolds in combination with Cho seeding may be successfully implemented for cartilage tissue engineering.

Scaffolding

In 2013, using a 3-d scaffolding of Matrigel in various configurations, substantial pancreatic organoids was produced in vitro. Clusters of small numbers of cells proliferated into 40,000 cells within one week. The clusters transform into cells that make either digestive enzymes or hormones like insulin, self-organizing into branched pancreatic organoids that resemble the pancreas.

The cells are sensitive to the environment, such as gel stiffness and contact with other cells. Individual cells do not thrive; a minimum of four proximate cells was required for subsequent organoid development. Modifications to the medium composition produced either hollow spheres mainly composed of pancreatic progenitors, or complex organoids that spontaneously undergo pancreatic morphogenesis and differentiation. Maintenance and expansion of pancreatic progenitors require active Notch and FGF signaling, recapitulating in vivo niche signaling interactions.

The organoids were seen as potentially offering mini-organs for drug testing and for spare insulin-producing cells.

Tissue Culture

Tissue culture is the growth of tissues or cells in an artificial medium separate from the organism. This is typically facilitated via use of a liquid, semi-solid, or solid growth medium, such as broth or agar. Tissue culture commonly refers to the culture of animal cells and tissues, with the more specific term plant tissue culture being used for plants. The term "tissue culture" was coined by American pathologist Montrose Thomas Burrows.

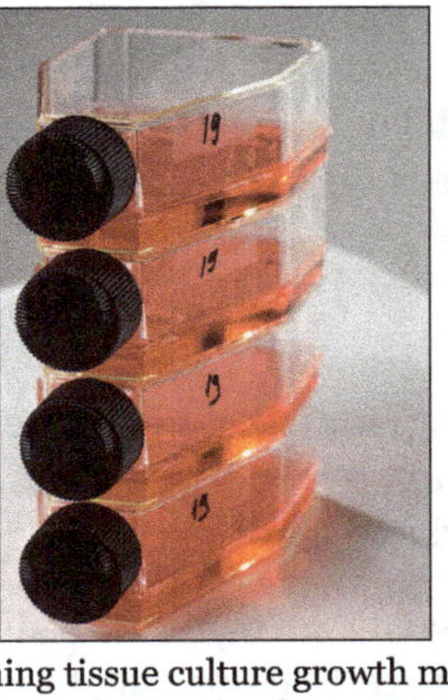

Flasks containing tissue culture growth medium which provides nourishment for the growing of cells.

In 1885 Wilhelm Roux removed a section of the medullary plate of an embryonic chicken and maintained it in a warm saline solution for several days, establishing the basic principle of tissue culture. In 1907 the zoologist Ross Granville Harrison demonstrated the growth of frog embryonic cells that would give rise to nerve cells in a medium of clotted lymph. In 1913, E. Steinhardt, C. Israeli, and R. A. Lambert grew vaccinia virus in fragments of guinea pig corneal tissue. In 1996, the first use of regenerative tissue was used to replace a small length of urethra, which led to the understanding that the technique of obtaining samples of tissue, growing it outside the body without a scaffold, and reapplying it, can be used for only small distances of less than 1 cm.

Gottlieb Haberlandt first pointed out the possibilities of the culture of isolated tissues, plant tissue culture. He suggested that the potentialities of individual cells via tissue culture as well as that the reciprocal influences of tissues on one another could be determined by this method. Since Haberlandt's original assertions, methods for tissue and cell culture have been realized, leading to significant discoveries in biology and medicine. His original idea, presented in 1902, was called totipotentiality: "Theoretically all plant cells are able to give rise to a complete plant."

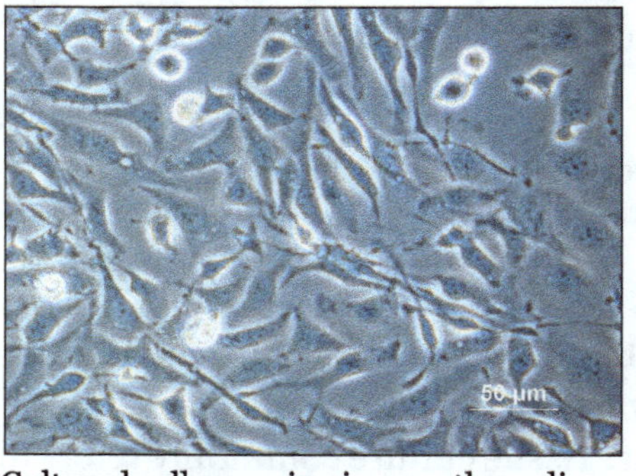

Cultured cells growing in growth medium.

In modern usage, tissue culture generally refers to the growth of cells from a tissue from a multicellular organism *in vitro*. These cells may be cells isolated from a donor organism, "primary cells", or an immortalised cell line. The cells are bathed in a culture medium, which contains essential nutrients and energy sources necessary for the cells' survival. The term tissue culture is often used interchangeably with cell culture.

The literal meaning of tissue culture refers to the culturing of tissue pieces, i.e. explant culture.

Tissue culture is an important tool for the study of the biology of cells from multicellular organisms. It provides an *in vitro* model of the tissue in a well defined environment which can be easily manipulated and analysed.

In animal tissue culture, cells may be grown as two-dimensional monolayers (conventional culture) or within fibrous scaffolds or gels to attain more naturalistic three-dimensional tissue-like structures (3D culture). Eric Simon, in a 1988 NIH SBIR grant report, showed that electrospinning could be used to produced nano- and submicron-scale polymeric fibrous scaffolds specifically intended for use as *in vitro* cell and tissue substrates. This early use of electrospun fibrous lattices for cell culture and tissue engineering showed that various cell types would adhere to and proliferate upon polycarbonate fibers. It was noted that as opposed to the flattened morphology typically seen in 2D culture, cells grown on the electrospun fibers exhibited a more rounded 3-dimensional morphology generally observed of tissues *in vivo*.

Plant tissue culture in particular is concerned with the growing of entire plants from small pieces of plant tissue, cultured in medium.

BIOMIMETICS

Biomimetics or biomimicry is the imitation of the models, systems, and elements of nature for the purpose of solving complex human problems.

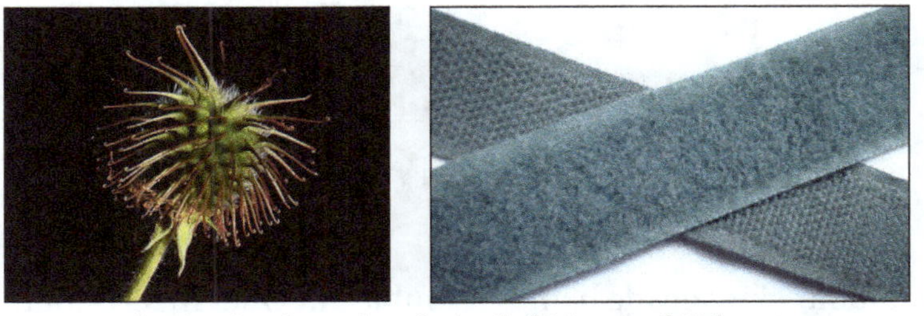

The tiny hooks on bur fruits (left) inspired Velcro tape.

Living organisms have evolved well-adapted structures and materials over geological time through natural selection. Biomimetics has given rise to new technologies inspired by biological solutions at macro and nanoscales. Humans have looked at nature for answers to problems throughout our existence. Nature has solved engineering problems such as self-healing abilities, environmental exposure tolerance and resistance, hydrophobicity, self-assembly, and harnessing solar energy.

Bio-inspired Technologies

Biomimetics could in principle be applied in many fields. Because of the diversity and complexity of biological systems, the number of features that might be imitated is large. Biomimetic applications are at various stages of development from technologies that might become commercially usable to prototypes. Murray's law, which in conventional form determined the optimum diameter of blood vessels, has been re-derived to provide simple equations for the pipe or tube diameter which gives a minimum mass engineering system.

Locomotion

Aircraft wing design and flight techniques are being inspired by birds and bats. Biorobots based on the physiology and methods of locomotion of animals include BionicKangaroo which moves like a kangaroo, saving energy from one jump and transferring it to its next jump. Kamigami Robots, a children's toy, mimic cockroach locomotion to run quickly and efficiently over indoor and outdoor surfaces.

Construction and Architecture

Researchers studied the termite's ability to maintain virtually constant temperature and humidity in their termite mounds in Africa despite outside temperatures that vary from 1.5 °C to 40 °C (35 °F to 104 °F). Researchers initially scanned a termite mound and created 3-D images of the mound structure, which revealed construction that could influence human building design. The Eastgate Centre, a mid-rise office complex in Harare, Zimbabwe, stays cool without air conditioning and uses only 10% of the energy of a conventional building of the same size.

A Waagner-Biro double-skin facade being assembled at One Angel Square, Manchester. The brown outer facade can be seen being assembled to the inner white facade via struts. These struts create a walkway between both 'skins' for ventilation, solar shading and maintenance

Researchers in the Sapienza University of Rome were inspired by the natural ventilation in termite mounds and designed a double façade that significantly cuts down over lit areas in a building. Scientists have imitated the porous nature of mound walls by designing a facade with double panels that was able to reduce heat gained by radiation and increase heat loss by convection in cavity between the two panels. The overall cooling load on the building's energy consumption was reduced by 15%.

A similar inspiration was drawn from the porous walls of termite mounds to design a naturally ventilated façade with a small ventilation gap. This design of façade is able to induce air flow due to the Venturi effect and continuously circulates rising air in the ventilation slot. Significant transfer of heat between the building's external wall surface and the air flowing over it was observed. The design is coupled with greening of the façade. Green wall facilitates additional natural cooling via evaporation, respiration and transpiration in plants.The damp plant substrate further support the cooling effect.

Scientists in Shanghai University were able to replicate the complex microstructure of clay-made conduit network in the mound to mimic the excellent humidity control in mounds. They proposed a porous humidity control material (HCM) using Sepiolite and calcium chloride with water vapor adsorption-desorption content at 550 grams per meter squared. Calcium chloride is a desiccant and improves the water vapor adsorption-desorption property of the Bio-HCM. The proposed bio-HCM has a regime of interfiber mesopores which acts as a mini reservoir. The flexural strength of the proposed material was estimated to be 10.3 MPa using computational simulations.

Sepiolite in solid form.

In structural engineering, the Swiss Federal Institute of Technology (EPFL) has incorporated biomimetic characteristics in an adaptive deployable "tensegrity" bridge. The bridge can carry out self-diagnosis and self-repair. The arrangement of leaves on a plant has been adapted for better solar power collection.

Analysis of the elastic deformation happening when a pollinisator lands on the sheath-like perch part of the flower *Strelitzia reginae* (known as Bird-of-Paradise flower) has inspired architects and scientists from the University of Freiburg and University of Stuttgart for the creation of hingeless shading systems that can react to their environment. These bio-inspired products is sold under the name Flectofin.

Other hingeless bioinspired system includes Flectofold. Flectofold has been inspired from the trapping system developed by the carnivorous plant *Aldrovanda vesiculosa.*

Structural Materials

There is a great need for new structural materials that are light weight but offer exceptional combinations of stiffness, strength, and toughness.

Such materials would need to be manufactured into bulk materials with complex shapes at high volume and low cost and would serve a variety of fields such as construction, transportation, energy storage and conversion. In a classic design problem, strength and toughness are more likely to be mutually exclusive i.e., strong materials are brittle and tough materials are weak. However, natural materials with complex and hierarchical material gradients that span from nano- to macro-scales are both strong and tough. Generally, most natural materials utilize limited chemical components but complex material architectures that give rise to exceptional mechanical properties. Understanding the highly diverse and multi functional biological materials and discovering approaches to replicate such structures will lead to advanced and more efficient technologies. Bone, nacre (abalone shell), teeth, the dactyl clubs of stomatopod shrimps and bamboo are great examples of damage tolerant materials. The exceptional resistance to fracture of bone is due to complex deformation and toughening mechanisms that operate at spanning different size scales - nanoscale structure of protein molecules to macroscopic physiological scale.

Nacre exhibits similar mechanical properties however with rather simpler structure. Nacre shows a brick and mortar llike structure with thick mineral layer (0.2~0.9-μm) of closely packed aragonite structures and thin organic matrix (~20-nm). While thin films and micrometer sized samples that mimic these structures are already produced, successful production of bulk biomimetic structural materials is yet to be realized. However, numerous processing techniques have been proposed for producing nacre like materials.

Electron microscopy image of a fractured surface of nacre.

Biomorphic mineralization is a technique that produces materials with morphologies and structures resembling those of natural living organisms by using bio-structures as templates for mineralization. Compared to other methods of material production, biomorphic mineralization is facile, environmentally benign and economic.

Freeze casting (Ice templating), an inexpensive method to mimic natural layered structures was employed by researchers at Lawrence Berkeley National Laboratory to create alumina-Al-Si and IT HAP-epoxy layered composites that match the mechanical properties of bone with an equivalent mineral/organic content. Various further studies also employed similar methods to produce high strength and high toughness composites involving a variety of constituent phases.

Recent studies demonstrated production of cohesive and self supporting macroscopic tissue constructs that mimic living tissues by printing tens of thousands of heterologous picoliter droplets in software-defined, 3D millimeter-scale geometries. Efforts are also taken up to mimic the design of nacre in artificial composite materials using fused deposition modelling and the helicoidal structures of stomatopod clubs in the fabrication of high performance carbon fiber-epoxy composites.

Various established and novel additive manufacturing technologies like PolyJet printing, direct ink writing, 3D magnetic printing, multi-material magnetically assisted 3D printing and magnetically-assisted slip casting have also been utilized to mimic the complex micro-scale architectures of natural materials and provide huge scope for future research.

Spider web silk is as strong as the Kevlar used in bulletproof vests. Engineers could in principle use such a material, if it could be reengineered to have a long enough life, for parachute lines, suspension bridge cables, artificial ligaments for medicine, and other purposes. The self-sharpening teeth of many animals have been copied to make better cutting tools. New ceramics that exhibit giant electret hysteresis have also been realized.

Self Healing Materials

In general in biological systems, self healing occurs via chemical signals released at the site of fracture which initiate a systemic response that transport repairing agents to the fracture site thereby promoting autonomic healing. To demonstrate the use of micro-vascular networks for autonomic healing, researchers developed a microvascular coating–substrate architecture that mimics human skin. Bio-inspired self-healing structural color hydrogels that maintain the stability of an inverse opal structure and its resultant structural colors were developed. A self-repairing membrane inspired by rapid self-sealing processes in plants was developed for inflatable light weight structures such as rubber boats or Tensairity constructions. The researchers applied a thin soft cellular polyurethane foam coating on the inside of a fabric substrate, which closes the crack if the membrane is punctured with a spike. Self-healing materials, polymers and composite materials capable of mending cracks have been produced based on biological materials.

Surfaces

Surfaces that recreate properties of shark skin are intended to enable more efficient movement through water. Efforts have been made to produce fabric that emulates shark skin.

Surface tension biomimetics are being researched for technologies such as hydrophobic or hydrophilic coatings and microactuators.

Adhesion

Wet Adhesion

Some amphibians, such as tree and torrent frogs and arboreal salamanders, are able to attach to and move over wet or even flooded environments without falling. This kind of organisms have toe pads which are permanently wetted by mucus secreted from glands that open into the channels between epidermal cells. They attach to mating surfaces by wet adhesion and they are capable of climbing on wet rocks even when water is flowing over the surface. Tire treads have also been inspired by the toe pads of tree frogs.

Marine mussels can stick easily and efficiently to surfaces underwater under the harsh conditions of the ocean. Mussels use strong filaments to adhere to rocks in the inter-tidal zones of wave-swept beaches, preventing them from being swept away in strong sea currents. Mussel foot proteins attach the filaments to rocks, boats and practically any surface in nature including other mussels. These proteins contain a mix of amino acid residues which has been adapted specifically for adhesive purposes. Researchers from the University of California Santa Barbara borrowed and simplified chemistries that the mussel foot uses to overcome this engineering challenge of wet adhesion to create copolyampholytes, and one-component adhesive systems with potential for employment in nanofabrication protocols. Other research has proposed adhesive glue from mussels.

Dry Adhesion

Leg attachment pads of several animals, including many insects (e.g. beetles and flies), spiders and lizards (e.g. geckos), are capable of attaching to a variety of surfaces and are used for locomotion, even on vertical walls or across ceilings. Attachment systems in these organisms have similar structures at their terminal elements of contact, known as setae. Such biological examples have offered inspiration in order to produce climbing robots, boots and tape. Synthetic setae have also been developed for the production of dry adhesives.

Optics

Biomimetic materials are gaining increasing attention in the field of optics and photonics. There are still little known bioinspired or biomimetic products involving the photonic properties of plants or animals. However, understanding how nature designed such optical materials from biological resources is worth pursuing and might lead to future commercial products.

Macr cast on a Petri dish (diameter: 3.5cm).

Inspiration from Fruits and Plants

For instance, the chiral self-assembly of cellulose inspired by the *Pollia condensata* berry has been exploited to make optically active films. Such films are made from cellulose which is a biodegradable and biobased resource obtained from wood or cotton. The structural colours can potentially be everlasting and have more vibrant colour than the ones obtained from chemical absorption of light. *Pollia condensata* is not the only fruit showing a structural coloured skin; iridescence is also found in berries of other species such as *Margaritaria nobilis*. These fruits show iridescent colors in the blue-green region of the visible spectrum which gives the fruit a strong metallic and shiny visual appearance. The structural colours come from the organisation of cellulose chains in the fruit's epicarp, a part of the fruit skin. Each cell of the epicarp is made of a multilayered envelope that behaves like a Bragg reflector. However, the light which is reflected from the skin of these fruits is not polarised unlike the one arising from man-made replicates obtained from the self-assembly of cellulose nanocrystals into helicoids, which only reflect left-handed circularly polarised light.

The fruit of *Elaeocarpus angustifolius* also show structural colour that come arises from the presence of specialised cells called iridosomes which have layered structures. Similar iridosomes have also been found in Delarbrea michieana fruits.

In plants, multi layer structures can be found either at the surface of the leaves (on top of the epidermis), such as in *Selaginella willdenowii* or within specialized intra-cellular organelles, the so-called iridoplasts, which are located inside the cells of the upper epidermis. For instance, the rain forest plants Begonia pavonina have iridoplasts located inside the epidermal cells.

Structural colours have also been found in several algae, such as in the red alga *Chondrus crispus* (Irish Moss).

Inspiration from Animals

Structural coloration produces the rainbow colours of soap bubbles, butterfly wings and many beetle scales. Phase-separation has been used to fabricate ultra-white scattering membranes from polymethylmethacrylate, mimicking the beetle *Cyphochilus*. LED lights can be designed to mimic the patterns of scales on fireflies' abdomens, improving their efficiency.

Vibrant blue color of *Morpho* butterfly due to structural coloration has been mimicked by a variety of technologies.

Morpho butterfly wings are structurally coloured to produce a vibrant blue that does not vary with angle. This effect can be mimicked by a variety of technologies. Lotus Cars claim to have developed a paint that mimics the *Morpho* butterfly's structural blue colour. In 2007, Qualcomm commercialised an interferometric modulator display technology, "Mirasol", using *Morpho*-like optical interference. In 2010, the dressmaker Donna

Sgro made a dress from Teijin Fibers' Morphotex, an undyed fabric woven from structurally coloured fibres, mimicking the microstructure of *Morpho* butterfly wing scales.

Canon Inc.'s SubWavelength structure Coating uses wedge-shaped structures the size of the wavelength of visible light. The wedge-shaped structures cause a continuously changing refractive index as light travels through the coating, significantly reducing lens flare. This imitates the structure of a moth's eye.

Agricultural Systems

Holistic planned grazing, using fencing and herders, seeks to restore grasslands by carefully planning movements of large herds of livestock to mimic the vast herds found in nature. The natural system being mimicked and used as a template is grazing animals concentrated by pack predators that must move on after eating, trampling, and manuring an area, and returning only after it has fully recovered. Developed by Allan Savory, who in turn was inspired by the work of André Voisin, this method of grazing holds tremendous potential in building soil, increasing biodiversity, reversing desertification, and mitigating global warming, similar to what occurred during the past 40 million years as the expansion of grass-grazer ecosystems built deep grassland soils, sequestering carbon and cooling the planet.

Permaculture is a set of design principles centered around whole systems thinking, simulating or directly utilizing the patterns and resilient features observed in natural ecosystems. It uses these principles in a growing number of fields from regenerative agriculture, rewilding, community, and organizational design and development.

Other Uses

Some air conditioning systems use biomimicry in their fans to increase airflow while reducing power consumption.

Technologists like Jas Johl have speculated that the functionality of vacuole cells could be used to design highly adaptable security systems. "The functionality of a vacuole, a biological structure that guards and promotes growth, illuminates the value of adaptability as a guiding principle for security." The functions and significance of vacuoles are fractal in nature, the organelle has no basic shape or size; its structure varies according to the requirements of the cell. Vacuoles not only isolate threats, contain what's necessary, export waste, maintain pressure-they also help the cell scale and grow. Johl argues these functions are necessary for any security system design. The 500 Series Shinkansen used biomimicry to reduce energy consumption and noise levels while increasing passenger comfort.

Other Technologies

Protein folding has been used to control material formation for self-assembled functional nanostructures. Polar bear fur has inspired the design of thermal collectors and clothing. The light refractive properties of the moth's eye has been studied to reduce the reflectivity of solar panels.

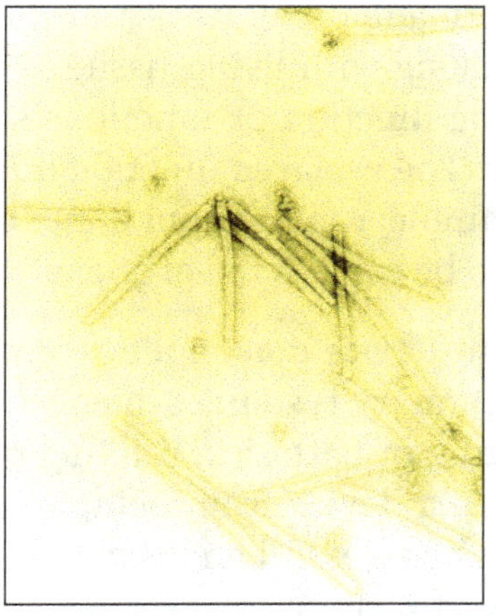

Scanning electron micrograph of rod shaped tobacco mosaic virus particles.

The Bombardier beetle's powerful repellent spray inspired a Swedish company to develop a "micro mist" spray technology, which is claimed to have a low carbon impact (compared to aerosol sprays). The beetle mixes chemicals and releases its spray via a steerable nozzle at the end of its abdomen, stinging and confusing the victim.

Most viruses have an outer capsule 20 to 300 nm in diameter. Virus capsules are remarkably robust and capable of withstanding temperatures as high as 60 °C; they are stable across the pH range 2-10. Viral capsules can be used to create nano device components such as nanowires, nanotubes, and quantum dots. Tubular virus particles such as the tobacco mosaic virus (TMV) can be used as templates to create nanofibers and nanotubes, since both the inner and outer layers of the virus are charged surfaces which can induce nucleation of crystal growth. This was demonstrated through the production of platinum and gold nanotubes using TMV as a template. Mineralized virus particles have been shown to withstand various pH values by mineralizing the viruses with different materials such as silicon, PbS, and CdS and could therefore serve as a useful carriers of material. A spherical plant virus called cowpea chlorotic mottle virus (CCMV) has interesting expanding properties when exposed to environments of pH higher than 6.5. Above this pH, 60 independent pores with diameters about 2 nm begin to exchange substance with the environment. The structural transition of the viral capsid can be utilized in Biomorphic mineralization for selective uptake and deposition of minerals by controlling the solution pH. Possible applications include using the viral cage to produce uniformly shaped and sized quantum dot semiconductor nanoparticles through a series of pH washes. This is an alternative to the apoferritin cage technique currently used to synthesize uniform CdSe nanoparticles. Such materials could also be used for targeted drug delivery since particles release contents upon exposure to specific pH levels.

ARTIFICIAL ORGAN

An artificial organ is a human made organ device or tissue that is implanted or integrated into a human — interfacing with living tissue — to replace a natural organ, to duplicate or augment a specific function or functions so the patient may return to a normal life as soon as possible. The replaced function does not have to be related to life support, but it often is. For example, replacement bones and joints, such as those found in hip replacements, could also be considered artificial organs.

Implied by definition, is that the device must not be continuously tethered to a stationary power supply or other stationary resources such as filters or chemical processing units. (Periodic rapid recharging of batteries, refilling of chemicals, and cleaning/replacing of filters would exclude a device from being called an artificial organ.) Thus, a dialysis machine, while a very successful and critically important life support device that almost completely replaces the duties of a kidney, is not an artificial organ.

Purpose

Constructing and installing artificial organs, an extremely research-intensive and expensive process initially, may entail many years of ongoing maintenance services not needed by a natural organ:

- Providing life support to prevent imminent death while awaiting a transplant (e.g. artificial heart).
- Dramatically improving the patient's ability for self care (e.g. artificial limb).
- Improving the patient's ability to interact socially (e.g. cochlear implant).
- Improving a patient's quality of life through cosmetic restoration after cancer surgery or an accident.

The use of any artificial organ by humans is almost always preceded by extensive experiments with animals. Initial testing in humans is frequently limited to those either already facing death or who have exhausted every other treatment possibility.

Artificial Limbs

Artificial arms and legs, or prosthetics, are intended to restore a degree of normal function to amputees. Mechanical devices that allow amputees to walk again or continue to use two hands have probably been in use since ancient times, the most notable one being the simple peg leg. Since then, the development of artificial limbs has progressed rapidly. New plastics and other materials, such as carbon fiber have allowed artificial limbs to become stronger and lighter, limiting the amount of extra energy necessary to operate the limb. Additional materials have allowed artificial limbs to look much more realistic. Prostheses can roughly be categorized as upper- and lower-extremity and can take many shapes and sizes.

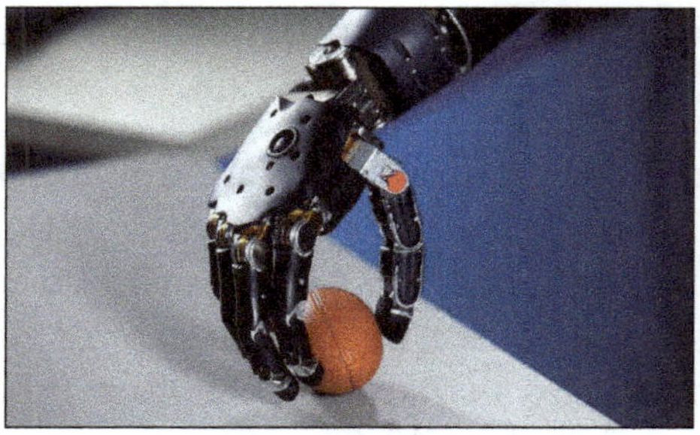

A prosthetic arm.

New advances in artificial limbs include additional levels of integration with the human body. Electrodes can be placed into nervous tissue, and the body can be trained to control the prosthesis. This technology has been used in both animals and humans. The prosthetic can be controlled by the brain using a direct implant or implant into various muscles.

Bladder

The two main methods for replacing bladder function involve either redirecting urine flow or replacing the bladder *in situ*. Standard methods for replacing the bladder involve fashioning a bladder-like pouch from intestinal tissue. As of 2017 methods to grow bladders using stem cells had been attempted in clinical research but this procedure was not part of medicine.

Brain

Neural prostheses are a series of devices that can substitute a motor, sensory or cognitive modality that might have been damaged as a result of an injury or a disease.

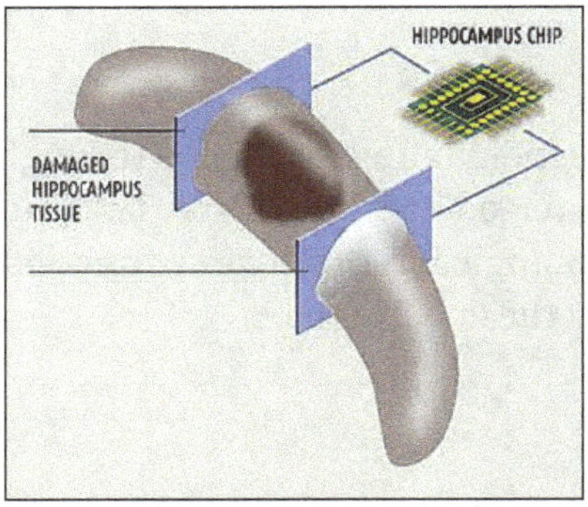

A diagram of a hippocampal prosthesis.

Neurostimulators, including deep brain stimulators, send electrical impulses to the brain in order to treat neurological and movement disorders, including Parkinson's disease, epilepsy, treatment resistant depression, and other conditions such as urinary incontinence. Rather than replacing existing neural networks to restore function, these devices often serve by disrupting the output of existing malfunctioning nerve centers to eliminate symptoms.

Scientists in 2013 created a mini brain that developed key neurological components until the early gestational stages of fetal maturation.

Scientists are also working on a project called mind uploading, which simply means uploading your "mind" o a computer chip or a data server. Data such as emotions, feelings, memories, abilities to think, and personalities will be collected and uploaded on a chip, that chip is supposed to function like a human brain, if the project is successful, human brains will be replaced by a small chip. Another thing, uploading minds on a data server is like creating a virtual world. This concept sounds fictional, and it is hard to understand. Human minds will be collected and uploaded on a virtual world, instead of living in a real world, they now all live in a world created by computer data, people might "live" in that world even when their real-life self are dead.

Elon Musk, current CEO of Tesla and SpaceX, wanted to connect human brains with computers. According to Musk, the project was started in 2020. If the project is a success, humans will be able to do everything they can with phones and computers just on their minds, simply thinking and computing inside their heads.

Corpora Cavernosa

To treat erectile dysfunction, both corpora cavernosa can be irreversibly surgically replaced with manually inflatable penile implants. This is a drastic therapeutic surgery meant only for men who suffer from complete impotence who have resisted all other treatment approaches. An implanted pump in the (groin) or (scrotum) can be manipulated by hand to fill these artificial cylinders, normally sized to be direct replacements for the natural corpora cavernosa, from an implanted reservoir in order to achieve an erection.

Testes

Men whom have sustained testicular abnormalities through birth defects or injury have been able to replace the damaged testicle with a testicular prosthesis. Although the prosthetic does not restore biological reproductive function, the device has been shown to improve mental health for these patients.

Ear

In cases when a person is profoundly deaf or severely hard of hearing in both ears, a cochlear implant may be surgically implanted. Cochlear implants bypass most of the peripheral auditory system to provide a sense of sound via a microphone and some electronics that reside outside the skin, generally behind the ear. The external components transmit a signal to an array of electrodes placed in the cochlea, which in turn stimulates the cochlear nerve. In the case of an outer ear trauma, a craniofacial prosthesis may be necessary.

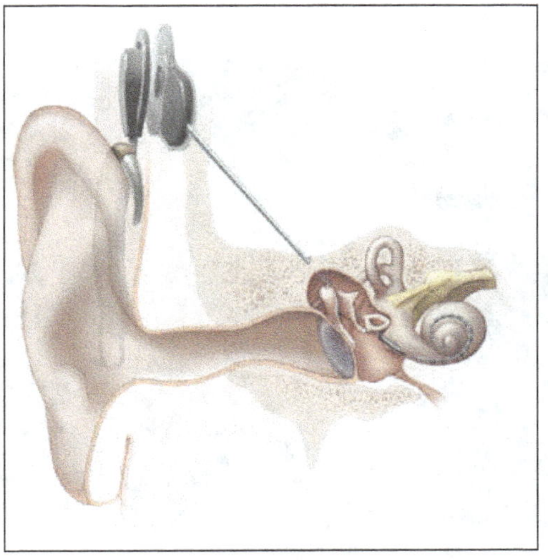

An illustration of a cochlear implant.

Eye

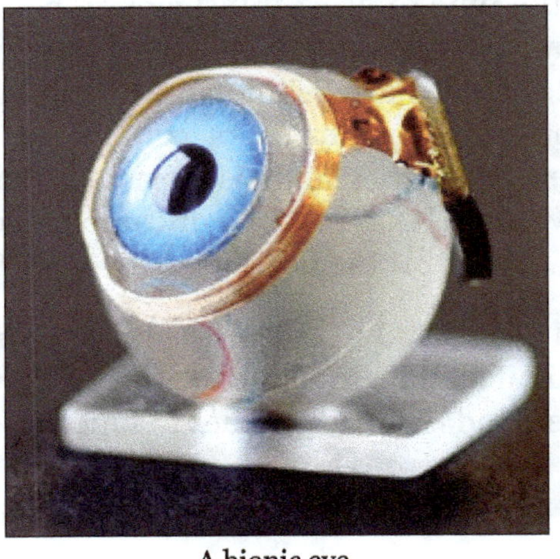

A bionic eye.

The most successful function-replacing artificial eye so far is actually an external miniature digital camera with a remote unidirectional electronic interface implanted on the retina, optic nerve, or other related locations inside the brain. The present state of the art yields only partial functionality, such as recognizing levels of brightness, swatches of color, and basic geometric shapes, proving the concept's potential.

Various researchers have demonstrated that the retina performs strategic image preprocessing for the brain. The problem of creating a completely functional artificial electronic eye is even more complex. Advances towards tackling the complexity of the artificial connection to the retina, optic nerve, or related brain areas, combined with ongoing advances in computer science, are expected to dramatically improve the performance of this technology.

Heart

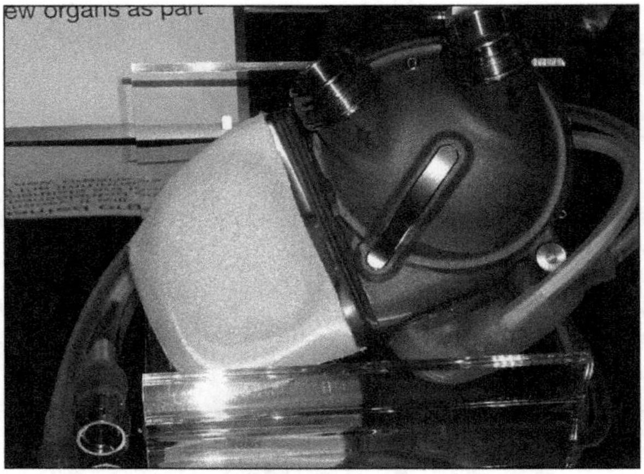

An artificial heart.

Cardiovascular-related artificial organs are implanted in cases where the heart, its valves, or another part of the circulatory system is in disorder. The artificial heart is typically used to bridge the time to heart transplantation, or to permanently replace the heart in case heart transplantation is impossible. Artificial pacemakers represent another cardiovascular device that can be implanted to either intermittently augment (defibrillator mode), continuously augment, or completely bypass the natural living cardiac pacemaker as needed. Ventricular assist devices are another alternative, acting as mechanical circulatory devices that partially or completely replace the function of a failing heart, without the removal of the heart itself.

Besides these, lab-grown hearts and 3D bioprinted hearts are also being researched. Currently, scientists are limited in their ability to grow and print hearts due to difficulties in getting blood vessels and lab-made tissues to function cohesively.

Kidney

It has been reported that scientists at the University of California, San Francisco, are developing an implantable artificial kidney. As of 2018, these scientists have made significant advancements with the technology but are still identifying methods to prevent the blood clotting associated with their machine.

The list of the patients who are waiting on kidneys are long, and kidneys are rare compared to other organs. Many people couldn't wait for their surgeries. Scientists feel the urge of developing an artificial kidney, they have been working hard in order to make a kidney that can function perfectly, and hopefully can replace human kidneys. Thanks to the NIBIB Quantum grantees, artificial kidney development advanced, they computed a simulation of how blood flow, they combined their work with a rare expertise in artificial kidney. "As developers of this technology know all too well, it is especially frustrating to deal with blood clots, which can both plug up the device, making in useless, and cause dangers to other parts of the body where blood flow would be compromised," said Rosemarie Hunziker, Director of the NIBIB program in Tissue Engineering and Regenerative Medicine.

An artificial kidney would allow blood filtrate continuously, that would help reduce kidney disease illness and increase the quality of life of patients.

Liver

HepaLife is developing a bioartificial liver device intended for the treatment of liver failure using stem cells. The artificial liver is designed to serve as a supportive device, either allowing the liver to regenerate upon failure, or to bridge the patient's liver functions until transplant is available. It is only made possible by the fact that it uses real liver cells (hepatocytes), and even then, it is not a permanent substitute.

Researchers from Japan found that a mixture of human liver precursor cells (differentiated from human induced pluripotent stem cells [iPSCs]) and two other cell types can spontaneously form three-dimensional structures dubbed "liver buds."

Lungs

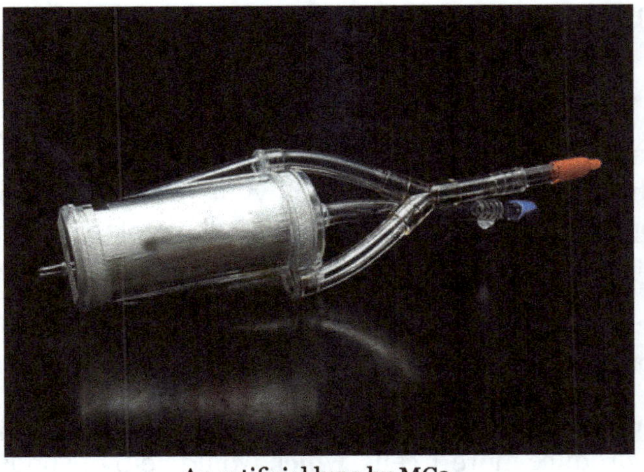

An artificial lung by MC3.

With some almost fully functional, artificial lungs promise to be a great success in the near future. An Ann Arbor company MC3 is currently working on this type of medical device.

Extracorporeal membrane oxygenation (ECMO) can be used to take significant load off of the native lung tissue and heart. In ECMO, a one or more catheters are placed into the patient and a pump is used to flow blood over hollow membrane fibers, which exchange oxygen and carbon dioxide with the blood. Similar to ECMO, Extracorporeal CO_2 Removal ($ECCO_2R$) has a similar set-up, but mainly benefits the patient through carbon dioxide removal, rather than oxygenation, with the goal of allowing the lungs to relax and heal.

Ovaries

The ground work for the development of the artificial ovary was laid in the early 1990s. Reproductive age patients who develop cancer often receive chemotherapy or radiation therapy, which damages oocytes and leads to early menopause. An artificial human ovary has been developed at Brown University with self-assembled microtissues created using novel 3-D petri dish technology. In a study funded and conducted by the NIH in 2017, scientists were successful in printing 3-D ovaries and implanting them in sterile mice. In the future, scientists hope to replicate this in larger animals as well as humans. The artificial ovary will be used for the purpose of in vitro maturation of immature oocytes and the development of a system to study the effect of environmental toxins on folliculogenesis.

Pancreas

An artificial pancreas is used to substitute endocrine functionality of a healthy pancreas for diabetic and other patients who require it. It can be used to improve insulin replacement therapy until glycemic control is practically normal as evident by the avoidance of the complications of hyperglycemia, and it can also ease the burden of therapy for the insulin-dependent. Approaches include using an insulin pump under closed loop control, developing a bio-artificial pancreas consisting of a biocompatible sheet of encapsulated beta cells, or using gene therapy.

Thymus

An implantable machine that performs the function of a thymus does not exist. However, researchers have been able to grow a thymus from reprogrammed fibroblasts. They expressed hope that the approach could one day replace or supplement neonatal thymus transplantation.

As of 2017, researchers at UCLA developed an artificial thymus that, although not yet implantable, is capable of performing all functions of a true thymus.

The artificial thymus would play an important role in the immune system, it would use blood stem cells to produce more T cells, which would help the body fight infections, it would also grant the body the ability to eliminate cancer cells. Since when people become old, their thymus don't work well, an artificial thymus would be a good choice to replace an old, not-functioning-well thymus.

The idea of using T cells to fight against infections has been around for a time, but until recently, the idea of using a T cell source, an artificial thymus is proposed. "We know that the key to creating a consistent and safe supply of cancer-fighting T cells would be to control the process in a way that deactivates all T cell receptors in the transplanted cells, except for the cancer-fighting receptors," Crooks said. The scientist also found that the T cells produced by the artificial thymus carried a diverse range of T cell receptors and worked similarly to the T cells produced by a normal thymus. Since they can work like human thymus, artificial thymus can supply a consistent amount of T cells to the body for the patients who are in need of treatments.

Trachea

The field of artificial tracheas went through a period of high interest and excitement with the work of Paolo Macchiarini at the Karolinska Institute and elsewhere from 2008 to around 2014, with front-page coverage in newspapers and on television. Concerns were raised about his work in 2014 and by 2016 he had been fired and high level management at Karolinska had been dismissed, including people involved in the Nobel Prize.

As of 2017 engineering a trachea—a hollow tube lined with cells—had proved more challenging then originally thought; challenges include the difficult clinical situation of people who present as clinical candidates, who generally have been through multiple procedures already; creating an implant that can become fully developed and integrate with host while withstanding respiratory forces, as well as the rotational and longitudinal movement the trachea undergoes.

Enhancement

It is also possible to construct and install an artificial organ to give its possessor abilities that are not naturally occurring. Research is proceeding in areas of vision, memory, and information processing. Some current research focuses on restoring short-term memory in accident victims and long-term memory in dementia patients.

One area of success was achieved when Kevin Warwick carried out a series of experiments extending his nervous system over the internet to control a robotic hand and the first direct electronic communication between the nervous systems of two humans.

This might also include the existing practice of implanting subcutaneous chips for identification and location purposes.

Microchips

Organ chips are devices containing hollow microvessels filled with cells simulating tissue and organs as a microfluidic system that can provide key chemical and electrical signal information.

This information can create various applications such as creating "human in vitro models" for both healthy and diseased organs, drug advancements in toxicity screening as well as replacing animal testing.

Using 3D cell culture techniques enables scientists to recreate the complex extracellular matrix, ECM, found in in vivo to mimic human response to drugs and human diseases. Organs on chips are used to reduce the failure rate in new drug development; microengineering these allows for a microenvironment to be modeled as an organ.

NEURAL ENGINEERING

Neural engineering (also known as neuroengineering) is a discipline within biomedical engineering that uses engineering techniques to understand, repair, replace, or enhance neural systems. Neural engineers are uniquely qualified to solve design problems at the interface of living neural tissue and non-living constructs.

The field of neural engineering draws on the fields of computational neuroscience, experimental neuroscience, clinical neurology, electrical engineering and signal processing of living neural tissue, and encompasses elements from robotics, cybernetics, computer engineering, neural tissue engineering, materials science, and nanotechnology.

Prominent goals in the field include restoration and augmentation of human function via direct interactions between the nervous system and artificial devices.

Much current research is focused on understanding the coding and processing of information in the sensory and motor systems, quantifying how this processing is altered in the pathological state, and how it can be manipulated through interactions with artificial devices including brain-computer interfaces and neuroprosthetics.

Other research concentrates more on investigation by experimentation, including the use of neural implants connected with external technology.

Neurohydrodynamics is a division of neural engineering that focuses on hydrodynamics of the neurological system.

Fundamentals

The fundamentals behind neuroengineering involve the relationship of neurons, neural networks, and nervous system functions to quantifiable models to aid the development of devices that could interpret and control signals and produce purposeful responses.

Neuroscience

Messages that the body uses to influence thoughts, senses, movements, and survival are directed by nerve impulses transmitted across brain tissue and to the rest of the body. Neurons are the basic functional unit of the nervous system and are highly specialized cells that are capable of sending these signals that operate high and low level functions needed for survival and quality of life. Neurons have special electro-chemical properties that allow them to process information and then transmit that information to other cells. Neuronal activity is dependent upon neural membrane potential and the changes that occur along and across it. A constant voltage, known as the membrane potential, is normally maintained by certain concentrations of specific ions across neuronal membranes. Disruptions or variations in this voltage create an imbalance, or polarization, across the membrane. Depolarization of the membrane past its threshold potential generates an action potential, which is the main source of signal transmission, known as neurotransmission of the nervous system. An action potential results in a cascade of ion flux down and across an axonal membrane, creating an effective voltage spike train or "electrical signal" which can transmit further electrical changes in other cells. Signals can be generated by electrical, chemical, magnetic, optical, and other forms of stimuli that influence the flow of charges, and thus voltage levels across neural membranes.

Engineering

Engineers employ quantitative tools that can be used for understanding and interacting with complex neural systems. Methods of studying and generating chemical, electrical, magnetic, and optical signals responsible for extracellular field potentials and synaptic transmission in neural tissue aid researchers in the modulation of neural system activity. To understand properties of neural system activity, engineers use signal processing techniques and computational modeling. To process these signals, neural engineers must translate the voltages across neural membranes into corresponding code, a process known as neural coding. Neural coding uses studies on how the brain encodes simple commands in the form of central pattern generators (CPGs), movement vectors, the cerebellar internal model, and somatotopic maps to understand movement and sensory phenomena. Decoding of these signals in the realm of neuroscience is the process by which neurons understand the voltages that have been transmitted to them. Transformations involve the mechanisms that signals of a certain form get interpreted and then translated into another form. Engineers look to mathematically model these transformations. There are a variety of methods being used to record these voltage signals. These can be intracellular or extracellular. Extracellular methods involve single-unit recordings, extracellular field potentials, and amperometry; more recently, multielectrode arrays have been used to record and mimic signals.

Scope

Neuromechanics

Neuromechanics is the coupling of neurobiology, biomechanics, sensation and perception, and robotics. Researchers are using advanced techniques and models to study the mechanical properties of neural tissues and their effects on the tissues' ability to withstand and generate force and movements as well as their vulnerability to traumatic loading. This area of research focuses on translating the transformations of information among the neuromuscular and skeletal systems to develop functions and governing rules relating to operation and organization of these systems. Neuromechanics can be simulated by connecting computational models of neural circuits to models of animal bodies situated in virtual physical worlds. Experimental analysis of biomechanics including the kinematics and dynamics of movements, the process and patterns of motor and sensory feedback during movement processes, and the circuit and synaptic organization of the brain responsible for motor control are all currently being researched to understand the complexity of animal movement. Dr. Michelle LaPlaca's lab at Georgia Institute of Technology is involved in the study of mechanical stretch of cell cultures, shear deformation of planar cell cultures, and shear deformation of 3D cell containing matrices. Understanding of these processes is followed by development of functioning models capable of characterizing these systems under closed loop conditions with specially defined parameters. The study of neuromechanics is aimed at improving treatments for physiological health problems which includes optimization of prostheses design, restoration of movement post injury, and design and control of mobile robots. By studying structures in 3D hydrogels, researchers can identify new models of nerve cell mechanoproperties. For example, LaPlaca et al. developed a new model showing that strain may play a role in cell culture.

Neuromodulation

Neuromodulation aims to treat disease or injury by employing medical device technologies that would enhance or suppress activity of the nervous system with the delivery of pharmaceutical agents, electrical signals, or other forms of energy stimulus to re-establish balance in impaired regions of the brain. Researchers in this field face the challenge of linking advances in understanding neural signals to advancements in technologies delivering and analyzing these signals with increased sensitivity, biocompatibility, and viability in closed loops schemes in the brain such that new treatments and clinical applications can be created to treat those suffering from neural damage of various kinds. Neuromodulator devices can correct nervous system dysfunction related to Parkinson's disease, dystonia, tremor, Tourette's, chronic pain, OCD, severe depression, and eventually epilepsy. Neuromodulation is appealing as treatment for varying defects because it focuses in on treating highly specific regions of the brain only, contrasting that of systemic treatments that can have side effects on the body. Neuromodulator stimulators such as microelectrode arrays can stimulate and record brain function and with further improvements are meant to become adjustable and responsive delivery devices for drugs and other stimuli.

Neural Regrowth and Repair

Neural engineering and rehabilitation applies neuroscience and engineering to investigating peripheral and central nervous system function and to finding clinical solutions to problems created by brain damage or malfunction. Engineering applied to neuroregeneration focuses on engineering devices and materials that facilitate the growth of neurons for specific applications such as the regeneration of peripheral nerve injury, the regeneration of the spinal cord tissue for spinal cord injury, and the regeneration of retinal tissue. Genetic engineering and tissue engineering are areas developing scaffolds for spinal cord to regrow across thus helping neurological problems.

Research and Applications

Research focused on neural engineering utilizes devices to study how the nervous system functions and malfunctions.

Neural Imaging

Neuroimaging techniques are used to investigate the activity of neural networks, as well as the structure and function of the brain. Neuroimaging technologies include functional magnetic resonance imaging (fMRI), magnetic resonance imaging (MRI), positron emission tomography (PET) and computed axial tomography (CAT) scans. Functional neuroimaging studies are interested in which areas of the brain perform specific tasks. fMRI measures hemodynamic activity that is closely linked to neural activity. It probes the brain by tuning the brain scanner to a certain wavelength to see which part of the brain are activated doing different tasks by seeing what lights up doing different things. PET, CT scanners, and electroencephalography (EEG) are currently being improved and used for similar purposes.

Neural Networks

Scientists can use experimental observations of neuronal systems and theoretical and computational models of these systems to create Neural networks with the hopes of modeling neural systems in as realistic a manner as possible. Neural networks can be used for analyses to help design further neurotechnological devices. Specifically, researchers handle analytical or finite element modeling to determine nervous system control of movements and apply these techniques to help patients with brain injuries or disorders. Artificial neural networks can be built from theoretical and computational models and implemented on computers from theoretically devices equations or experimental results of observed behavior of neuronal systems. Models might represent ion concentration dynamics, channel kinetics, synaptic transmission, single neuron computation, oxygen metabolism, or application of dynamic system theory. Liquid-based template assembly was used to engineer 3D neural networks from neuron-seeded microcarrier beads.

Neural Interfaces

Neural interfaces are a major element used for studying neural systems and enhancing or replacing neuronal function with engineered devices. Engineers are challenged with developing electrodes that can selectively record from associated electronic circuits to collect information about the nervous system activity and to stimulate specified regions of neural tissue to restore function or sensation of that tissue. The materials used for these devices must match the mechanical properties of neural tissue in which they are placed and the damage must be assessed. Neural interfacing involves temporary regeneration of biomaterial scaffolds or chronic electrodes and must manage the body's response to foreign materials. Microelectrode arrays are recent advances that can be used to study neural networks. Optical neural interfaces involve optical recordings and optogenetics stimulation that makes brain cells light sensitive. Fiber optics can be implanted in the brain to stimulate and record this photon activity instead of electrodes. Two-photon excitation microscopy can study living neuronal networks and the communicatory events among neurons.

Brain Computer Interfaces

Brain computer interfaces seek to directly communicate with human nervous system to monitor and stimulate neural circuits as well as diagnose and treat intrinsic neurological dysfunction. Deep brain stimulation is a significant advance in this field that is especially effective in treating movement disorders such as Parkinson's disease with high frequency stimulation of neural tissue to suppress tremors.

Microsystems

Neural microsystems can be developed to interpret and deliver electrical, chemical, magnetic, and optical signals to neural tissue. They can detect variations in membrane potential and measure electrical properties such as spike population, amplitude, or rate by using electrodes, or by assessment of chemical concentrations, fluorescence light intensity, or magnetic field potential. The goal of these systems is to deliver signals that would influence neuronal tissue potential and thus stimulate the brain tissue to evoke a desired response.

Microelectrode Arrays

Microelectrode arrays are specific tools used to detect the sharp changes in voltage in the extracellular environments that occur from propagation of an action potential down an axon. Dr. Mark Allen and Dr. LaPlaca have microfabricated 3D electrodes out of cytocompatible materials such as SU-8 and SLA polymers which have led to the development of in vitro and in vivo microelectrode systems with the characteristics of high compliance and flexibility to minimize tissue disruption.

Neural Prostheses

Neuroprosthetics are devices capable of supplementing or replacing missing functions of the nervous system by stimulating the nervous system and recording its activity. Electrodes that measure firing of nerves can integrate with prosthetic devices and signal them to perform the function intended by the transmitted signal. Sensory prostheses use artificial sensors to replace neural input that might be missing from biological sources. Engineers researching these devices are charged with providing a chronic, safe, artificial interface with neuronal tissue. Perhaps the most successful of these sensory prostheses is the cochlear implant which has restored hearing abilities to the deaf. Visual prosthesis for restoring visual capabilities of blind persons is still in more elementary stages of development. Motor prosthesics are devices involved with electrical stimulation of biological neural muscular system that can substitute for control mechanisms of the brain or spinal cord. Smart prostheses can be designed to replace missing limbs controlled by neural signals by transplanting nerves from the stump of an amputee to muscles. Sensory prosthetics provide sensory feedback by transforming mechanical stimuli from the periphery into encoded information accessible by the nervous system. Electrodes placed on the skin can interpret signals and then control the prosthetic limb. These prosthetics have been very successful. Functional electrical stimulation (FES) is a system aimed at restoring motor processes such as standing, walking, and hand grasp.

Neurorobotics

Neurorobotics is the study of how neural systems can be embodied and movements emulated in mechanical machines. Neurorobots are typically used to study motor control and locomotion, learning and memory selection, and value systems and action selection. By studying neurorobots in real-world environments, they are more easily observed and assessed to describe heuristics of robot function in terms of its embedded neural systems and the reactions of these systems to its environment. For instance, making use of a computational model of epilectic spike-wave dynamics, it has been already proven the effectiveness of a method to simulate seizure abatement through a pseudospectral protocol. The computational model emulates the brain connectivity by using a magnetic imaging resonance from a patient suffering of idiopathic generalized epilepsy. The method was able to generate stimuli able to lessen the seizures.

Neural Tissue Regeneration

Neural tissue regeneration, or neuroregeneration looks to restore function to those neurons that have been damaged in small injuries and larger injuries like those caused by traumatic brain injury. Functional restoration of damaged nerves involves re-establishment of a continuous pathway for regenerating axons to the site of innervation. Researchers like Dr. LaPlaca at Georgia Institute of Technology are looking to help find treatment for repair and regeneration after traumatic brain injury and spinal cord injuries by applying tissue engineering strategies. Dr. LaPlaca is looking into methods combining neural stem cells with an extracellular matrix protein based scaffold for minimally invasive delivery into the irregular shaped lesions that form after a traumatic insult. By studying the neural stem cells in vitro and exploring alternative cell sources, engineering novel biopolymers that could be utilized in a scaffold, and investigating cell or tissue engineered construct transplants in vivo in models of traumatic brain and spinal cord injury, Dr. LaPlaca's lab aims to identify optimal strategies for nerve regeneration post injury.

Current Approaches to Clinical Treatment

End to end surgical suture of damaged nerve ends can repair small gaps with autologous nerve grafts. For larger injuries, an autologous nerve graft that has been harvested from another site in the body might be used, though this process is time consuming, costly and requires two surgeries. Clinical treatment for CNS is minimally available and focuses mostly on reducing collateral damage caused by bone fragments near the site of injury or inflammation. After swelling surrounding injury lessens, patients undergo rehabilitation so that remaining nerves can be trained to compensate for the lack of nerve function in injured nerves. No treatment currently exists to restore nerve function of CNS nerves that have been damaged.

Engineering Strategies for Repair

Engineering strategies for the repair of spinal cord injury are focused on creating a friendly environment for nerve regeneration. Only PNS nerve damage has been clinically possible so far, but advances in research of genetic techniques and biomaterials demonstrate the potential for SC nerves to regenerate in permissible environments.

Grafts

Advantages of autologous tissue grafts are that they come from natural materials which have a high likelihood of biocompatibility while providing structural support to nerves that encourage cell adhesion and migration. Nonautologous tissue, acellular grafts, and extracellular matrix based materials are all options that may also provide ideal scaffolding for nerve regeneration. Some come from allogenic or xenogenic tissues that must be combined with immunosuppressants. while others include small intestinal submucosa and amniotic tissue grafts. Synthetic materials are attractive options because their physical and chemical properties can typically be controlled. A challenge that remains with synthetic materials is biocompatibility. Methylcellulose-based constructs have been shown to be a biocompatible option serving this purpose. AxoGen uses a cell graft technology AVANCE to mimic a human nerve. It has been shown to achieve meaningful recovery in 87 percent of patients with peripheral nerve injuries.

Nerve Guidance Channels

Nerve guidance channels, Nerve guidance conduit are innovative strategies focusing on larger defects that provide a conduit for sprouting axons directing growth and reducing growth inhibition from scar tissue. Nerve guidance channels must be readily formed into a conduit with the desired dimensions, sterilizable, tear resistant, and easy to handle and suture. Ideally they would degrade over time with nerve regeneration, be pliable, semipermeable, maintain their shape, and have a smooth inner wall that mimics that of a real nerve.

Biomolecular Therapies

Highly controlled delivery systems are needed to promote neural regeneration. Neurotrophic factors can influence development, survival, outgrowth, and branching. Neurotrophins include nerve growth factor (NGF), brain derived neurotrophic factor (BDNF), neurotrophin-3 (NT-3) and neurotrophin-4/5 (NT-4/5). Other factors are ciliary neurotrophic factor (CNTF), glial cell line-derived growth factor (GDNF) and acidic and basic fibroblast growth factor (aFGF, bFGF) that promote a range of neural responses. Fibronectin has also been shown to support nerve regeneration following TBI in rats. Other therapies are looking into regeneration of nerves by upregulating regeneration associated genes (RAGs), neuronal cytoskeletal components, and antiapoptosis factors. RAGs include GAP-43 and Cap-23, adhesion molecules such as L1 family, NCAM, and N-cadherin. There is also the potential for blocking inhibitory biomolecules in the CNS due to glial scarring. Some currently being studied are treatments with chondroitinase ABC and blocking NgR, ADP-ribose.

Delivery Techniques

Delivery devices must be biocompatible and stable in vivo. Some examples include osmotic pumps, silicone reservoirs, polymer matrices, and microspheres. Gene therapy techniques have also been studied to provide long-term production of growth factors and could be delivered with viral or non-viral vectors such as lipoplexes. Cells are also effective delivery vehicles for ECM components, neurotrophic factors and cell adhesion molecules. Olfactory ensheathing cells (OECs) and stem cells as well as genetically modified cells have been used as transplants to support nerve regeneration.

Advanced Therapies

Advanced therapies combine complex guidance channels and multiple stimuli that focus on internal structures that mimic the nerve architecture containing internal matrices of longitudinally aligned fibers or channels. Fabrication of these structures can use a number of technologies: magnetic polymer fiber alignment, injection molding, phase separation, solid free-form fabrication, and ink jet polymer printing.

Neural Enhancement

Augmentation of human neural systems, or human enhancement using engineering techniques is another possible application of neuroengineering. Deep brain stimulation has already been shown to enhance memory recall as noted by patients currently using this treatment for neurological disorders. Brain stimulation techniques are postulated to be able to sculpt emotions and personalities as well as enhance motivation, reduce inhibitions, etc. as requested by the individual. Ethical issues with this sort of human augmentation are a new set of questions that neural engineers have to grapple with as these studies develop.

PHARMACEUTICAL ENGINEERING

Pharmaceutical engineering is a branch of engineering focused on discovering, formulating, and manufacturing medication, as well as analytical and quality control processes.

It utilizes the fields of chemical engineering, biomedical engineering, and pharmaceutical sciences.

Humans have a long history of using derivatives of natural resources, such as plants, as medication. However, it was not until the late 19th century when the technological advancements of chemical companies were combined with medical research that scientists began to manipulate and engineer new medications, drug delivery techniques, and methods of mass production.

Synthesizing New Medications

One of the first prominent examples of an engineered, synthetic medication was made by Paul Erlich. Erlich had found that Atoxyl, an arsenic-containing compound which is harmful to humans, was very effective at killing Treponema pallidum, the bacteria which causes Syphilis. He hypothesized that if the structure of Atoxyl was altered, a "magic bullet" could potentially be identified which would kill the parasitic bacteria without having any adverse effects on human health. He developed many compounds stemming from the chemical structure of Atoxyl and eventually identified one compound which was the most effective against Syphilis while being the least harmful to humans, which became known as Salvarsan. Salvarsan was widely used to treat Syphilis within years of its discovery.

Beginning of Mass Production

Equipment for deep-fermentation of penicillin.

In 1928, Alexander Fleming discovered a mold named Penicillium chrysogenum which prevented many types of bacteria from growing. Scientists identified the potential of this mold to provide treatment in humans against bacteria which cause infections. During World War II, the United Kingdom and the United States worked together to find a method of mass producing Penicillin, a derivative of the Penicillium mold, which had the potential to save many lives during the war since it could treat infections common in injured soldiers. Although Penicillin could be isolated from the mold in a laboratory setting, there was no known way to obtain the amount of medication needed to treat the quantity of people who needed it. Scientists with major chemical companies such as Pfizer were able to develop a deep-fermentation process which could produce a high yield of penicillin. In 1944, Pfizer opened the first penicillin factory, and its products were exported to aid the war efforts overseas.

Controlled Drug Release

Tablets for oral consumption of medication have been utilized since approximately 1500 B.C., however for a long time the only method of drug release was immediate release, meaning all of the medication is released in the body at once. In the 1950s, sustained release technology was developed. Through mechanisms such as osmosis and diffusion, pills were designed that could release the medication over a 12-hour to 24-hour period. Smith, Kline & French developed one of the first major successful sustained release technologies. Their formulation consisted of a collection of small tablets taken at the same time, with varying amounts of wax coating that allowed some tablets to dissolve in the body faster than others. The result was a continuous release of the drug as it travelled through the intestinal tract. Although modern day research focuses on extending the controlled release timescale to the order of months, once-a-day and twice-a-day pills are still the most widely utilized controlled drug release method.

Formation of the ISPE

In 1980, the International Society for Pharmaceutical Engineering was formed to support and guide professionals in the pharmaceutical industry through all parts of the process of bringing new medications to the market. The ISPE writes standards and guidelines for individuals and companies to use and to model their practices after. The ISPE also hosts training sessions and conferences for professionals to attend, learn, and collaborate with others in the field.

BIOINFORMATICS

Bioinformatics is an interdisciplinary field that develops methods and software tools for understanding biological data. As an interdisciplinary field of science, bioinformatics combines biology, computer science, information engineering, mathematics and statistics to analyze and interpret biological data. Bioinformatics has been used for *in silico* analyses of biological queries using mathematical and statistical techniques.

Bioinformatics includes biological studies that use computer programming as part of their methodology, as well as a specific analysis "pipelines" that are repeatedly used, particularly in the field of genomics. Common uses of bioinformatics include the identification of candidates genes and single nucleotide polymorphisms (SNPs). Often, such identification is made with the aim of better understanding the genetic basis of disease, unique adaptations, desirable properties (esp. in agricultural species), or differences between populations. In a less formal way, bioinformatics also tries to understand the organisational principles within nucleic acid and protein sequences, called proteomics.

Sequence Analysis

Since the Phage Φ-X174 was sequenced in 1977, the DNA sequences of thousands of organisms have been decoded and stored in databases. This sequence information is analyzed to determine genes that encode proteins, RNA genes, regulatory sequences, structural motifs, and repetitive sequences. A comparison of genes within a species or between different species can show similarities between protein functions, or relations between species (the use of molecular systematics to construct phylogenetic trees). With the growing amount of data, it long ago became impractical to analyze DNA sequences manually. Today, computer programs such as BLAST are used daily to search sequences from more than 260 000 organisms, containing over 190 billion nucleotides. These programs can compensate for mutations (exchanged, deleted or inserted bases) in the DNA sequence, to identify sequences that are related, but not identical. A variant of this sequence alignment is used in the sequencing process itself. For the special task of taxonomic classification of sequence snippets, modern k-mer based software like Kraken achieves throughput unreachable by alignment methods.

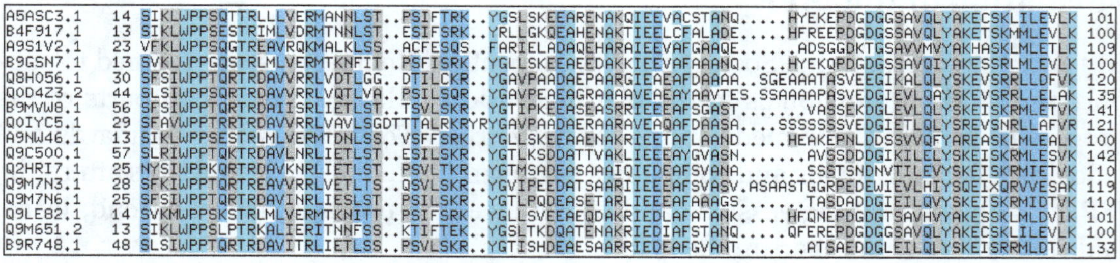

The sequences of different genes or proteins may be aligned side-by-side to measure their similarity. This alignment compares protein sequences and genomic sequences containing WPP domains.

DNA Sequencing

Before sequences can be analyzed they have to be obtained from the data storage bank example the Genbank. DNA sequencing is still a non-trivial problem as the raw data may be noisy or afflicted by weak signals. Algorithms have been developed for base calling for the various experimental approaches to DNA sequencing.

Sequence Assembly

Most DNA sequencing techniques produce short fragments of sequence that need to be assembled to obtain complete gene or genome sequences. The so-called shotgun sequencing technique (which was used, for example, by The Institute for Genomic Research (TIGR) to sequence the first bacterial genome, *Haemophilus influenzae*) generates the sequences of many thousands of small DNA fragments (ranging from 35 to 900 nucleotides long, depending on the sequencing technology). The ends of these fragments overlap and, when aligned properly by a genome assembly program, can be used to reconstruct the complete genome. Shotgun sequencing yields sequence data quickly, but the task of assembling the fragments can be quite complicated for larger genomes. For a genome as large as the human genome, it may take many days of CPU time on large-memory, multiprocessor computers to assemble the fragments, and the resulting assembly usually contains numerous gaps that must be filled in later. Shotgun sequencing is the method of choice for virtually all genomes sequenced today, and genome assembly algorithms are a critical area of bioinformatics research.

Genome Annotation

In the context of genomics, annotation is the process of marking the genes and other biological features in a DNA sequence. This process needs to be automated because most genomes are too large to annotate by hand, not to mention the desire to annotate as many genomes as possible, as the rate of sequencing has ceased to pose a bottleneck. Annotation is made possible by the fact that genes have recognisable start and stop regions, although the exact sequence found in these regions can vary between genes.

The first description of a comprehensive genome annotation system was published in 1995 by the team at The Institute for Genomic Research that performed the first complete sequencing and analysis of the genome of a free-living organism, the bacterium *Haemophilus influenzae*. Owen White designed and built a software system to identify the genes encoding all proteins, transfer RNAs, ribosomal RNAs (and other sites) and to make initial functional assignments. Most current genome annotation systems work similarly, but the programs available for analysis of genomic DNA, such as the GeneMark program trained and used to find protein-coding genes in *Haemophilus influenzae*, are constantly changing and improving.

Following the goals that the Human Genome Project left to achieve after its closure in 2003, a new project developed by the National Human Genome Research Institute in the U.S appeared. The so-called ENCODE project is a collaborative data collection of the functional elements of the human genome that uses next-generation DNA-sequencing technologies and genomic tiling arrays, technologies able to automatically generate large amounts of data at a dramatically reduced per-base cost but with the same accuracy (base call error) and fidelity (assembly error).

Computational Evolutionary Biology

Evolutionary biology is the study of the origin and descent of species, as well as their change over time. Informatics has assisted evolutionary biologists by enabling researchers to:

- Trace the evolution of a large number of organisms by measuring changes in their DNA, rather than through physical taxonomy or physiological observations alone.

- Compare entire genomes, which permits the study of more complex evolutionary events, such as gene duplication, horizontal gene transfer, and the prediction of factors important in bacterial speciation.
- Build complex computational population genetics models to predict the outcome of the system over time.
- Track and share information on an increasingly large number of species and organisms.

Future work endeavours to reconstruct the now more complex tree of life.

Comparative Genomics

The core of comparative genome analysis is the establishment of the correspondence between genes (orthology analysis) or other genomic features in different organisms. It is these intergenomic maps that make it possible to trace the evolutionary processes responsible for the divergence of two genomes. A multitude of evolutionary events acting at various organizational levels shape genome evolution. At the lowest level, point mutations affect individual nucleotides. At a higher level, large chromosomal segments undergo duplication, lateral transfer, inversion, transposition, deletion and insertion. Ultimately, whole genomes are involved in processes of hybridization, polyploidization and endosymbiosis, often leading to rapid speciation. The complexity of genome evolution poses many exciting challenges to developers of mathematical models and algorithms, who have recourse to a spectrum of algorithmic, statistical and mathematical techniques, ranging from exact, heuristics, fixed parameter and approximation algorithms for problems based on parsimony models to Markov chain Monte Carlo algorithms for Bayesian analysis of problems based on probabilistic models.

Many of these studies are based on the detection of sequence homology to assign sequences to protein families.

Pan Genomics

Pan genomics is a concept introduced in 2005 by Tettelin and Medini which eventually took root in bioinformatics. Pan genome is the complete gene repertoire of a particular taxonomic group: although initially applied to closely related strains of a species, it can be applied to a larger context like genus, phylum etc. It is divided in two parts- The Core genome: Set of genes common to all the genomes under study (These are often housekeeping genes vital for survival) and The Dispensable/Flexible Genome: Set of genes not present in all but one or some genomes under study. A bioinformatics tool BPGA can be used to characterize the Pan Genome of bacterial species.

Genetics of Disease

With the advent of next-generation sequencing we are obtaining enough sequence data to map the genes of complex diseases infertility, breast cancer or Alzheimer's disease. Genome-wide association studies are a useful approach to pinpoint the mutations responsible for such complex diseases. Through these studies, thousands of DNA variants have been identified that are associated with similar diseases and traits. Furthermore, the possibility for genes to be used at prognosis, diagnosis or treatment is one of the most essential applications. Many studies are discussing both the promising ways to choose the genes to be used and the problems and pitfalls of using genes to predict disease presence or prognosis.

Analysis of Mutations in Cancer

In cancer, the genomes of affected cells are rearranged in complex or even unpredictable ways. Massive sequencing efforts are used to identify previously unknown point mutations in a variety of genes in cancer. Bioinformaticians continue to produce specialized automated systems to manage the sheer volume of sequence data produced, and they create new algorithms and software to compare the sequencing results to the growing collection of human genome sequences and germline polymorphisms. New physical detection technologies are employed, such as oligonucleotide microarrays to identify chromosomal gains and losses (called comparative genomic hybridization), and single-nucleotide polymorphism arrays to detect known *point mutations*. These detection methods simultaneously measure several hundred thousand sites throughout the genome, and when used in high-throughput to measure thousands of samples, generate terabytes of data per experiment. Again the massive amounts and new types of data generate new opportunities for bioinformaticians. The data is often found to contain considerable variability, or noise, and thus Hidden Markov model and change-point analysis methods are being developed to infer real copy number changes.

Two important principles can be used in the analysis of cancer genomes bioinformatically pertaining to the identification of mutations in the exome. First, cancer is a disease of accumulated somatic mutations in genes. Second cancer contains driver mutations which need to be distinguished from passengers.

With the breakthroughs that this next-generation sequencing technology is providing to the field of Bioinformatics, cancer genomics could drastically change. These new methods and software allow bioinformaticians to sequence many cancer genomes quickly and affordably. This could create a more flexible process for classifying types of cancer by analysis of cancer driven mutations in the genome. Furthermore, tracking of patients while the disease progresses may be possible in the future with the sequence of cancer samples.

Another type of data that requires novel informatics development is the analysis of lesions found to be recurrent among many tumors.

Gene and Protein Expression

Analysis of Gene Expression

The expression of many genes can be determined by measuring mRNA levels with multiple techniques including microarrays, expressed cDNA sequence tag (EST) sequencing, serial analysis of gene expression (SAGE) tag sequencing, massively parallel signature sequencing (MPSS), RNA-Seq, also known as "Whole Transcriptome Shotgun Sequencing" (WTSS), or various applications of multiplexed in-situ hybridization. All of these techniques are extremely noise-prone and subject to bias in the biological measurement, and a major research area in computational biology involves developing statistical tools to separate signal from noise in high-throughput gene expression studies. Such studies are often used to determine the genes implicated in a disorder: one might compare microarray data from cancerous epithelial cells to data from non-cancerous cells to determine the transcripts that are up-regulated and down-regulated in a particular population of cancer cells.

Analysis of Protein Expression

Protein microarrays and high throughput (HT) mass spectrometry (MS) can provide a snapshot of the proteins present in a biological sample. Bioinformatics is very much involved in making sense of protein microarray and HT MS data; the former approach faces similar problems as with microarrays targeted at mRNA, the latter involves the problem of matching large amounts of mass data against predicted masses from protein sequence databases, and the complicated statistical analysis of samples where multiple, but incomplete peptides from each protein are detected. Cellular protein localization in a tissue context can be achieved through affinity proteomics displayed as spatial data based on immunohistochemistry and tissue microarrays.

Analysis of Regulation

Gene regulation is the complex orchestration of events by which a signal, potentially an extracellular signal such as a hormone, eventually leads to an increase or decrease in the activity of one or more proteins. Bioinformatics techniques have been applied to explore various steps in this process.

For example, gene expression can be regulated by nearby elements in the genome. Promoter analysis involves the identification and study of sequence motifs in the DNA surrounding the coding region of a gene. These motifs influence the extent to which that region is transcribed into mRNA. Enhancer elements far away from the promoter can also regulate gene expression, through three-dimensional looping interactions. These interactions can be determined by bioinformatic analysis of chromosome conformation capture experiments.

Expression data can be used to infer gene regulation: one might compare microarray data from a wide variety of states of an organism to form hypotheses about the genes involved in each state. In a single-cell organism, one might compare stages of the cell cycle, along with various stress conditions (heat shock, starvation, etc.). One can then apply clustering algorithms to that expression data to determine which genes are co-expressed. For example, the upstream regions (promoters) of co-expressed genes can be searched for over-represented regulatory elements. Examples of clustering algorithms applied in gene clustering are k-means clustering, self-organizing maps (SOMs), hierarchical clustering, and consensus clustering methods.

Analysis of Cellular Organization

Several approaches have been developed to analyze the location of organelles, genes, proteins, and other components within cells. This is relevant as the location of these components affects the events within a cell and thus helps us to predict the behavior of biological systems. A gene ontology category, *cellular compartment*, has been devised to capture subcellular localization in many biological databases.

Microscopy and Image Analysis

Microscopic pictures allow us to locate both organelles as well as molecules. It may also help us to distinguish between normal and abnormal cells, e.g. in cancer.

Protein Localization

The localization of proteins helps us to evaluate the role of a protein. For instance, if a protein is found in the nucleus it may be involved in gene regulation or splicing. By contrast, if a protein is found in mitochondria, it may be involved in respiration or other metabolic processes. Protein localization is thus an important component of protein function prediction. There are well developed protein subcellular localization prediction resources available, including protein subcellular location databases, and prediction tools.

Nuclear Organization of Chromatin

Data from high-throughput chromosome conformation capture experiments, such as Hi-C (experiment) and ChIA-PET, can provide information on the spatial proximity of DNA loci. Analysis of these experiments can determine the three-dimensional structure and nuclear organization of chromatin. Bioinformatic challenges in this field include partitioning the genome into domains, such as Topologically Associating Domains (TADs), that are organised together in three-dimensional space.

Structural Bioinformatics

Protein structure prediction is another important application of bioinformatics. The amino acid sequence of a protein, the so-called primary structure, can be easily determined from the sequence on the gene that codes for it. In the vast majority of cases, this primary structure uniquely determines a structure in its native environment. (Of course, there are exceptions, such as the bovine spongiform encephalopathy (mad cow disease) prion.) Knowledge of this structure is vital in understanding the function of the protein. Structural information is usually classified as one of *secondary*, *tertiary* and *quaternary* structure. A viable general solution to such predictions remains an open problem. Most efforts have so far been directed towards heuristics that work most of the time.

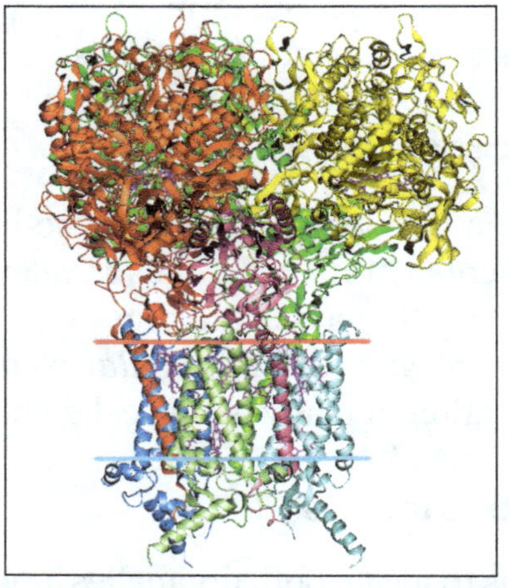

3-dimensional protein structures such as this one are common subjects in bioinformatic analyses.

One of the key ideas in bioinformatics is the notion of homology. In the genomic branch of bioinformatics, homology is used to predict the function of a gene: if the sequence of gene *A*, whose function is known, is homologous to the sequence of gene *B*, whose function is unknown, one could infer that B may share A's function. In the structural branch of bioinformatics, homology is used to determine which parts of a protein are important in structure formation and interaction with other proteins. In a technique called homology modeling, this information is used to predict the structure of a protein once the structure of a homologous protein is known. This currently remains the only way to predict protein structures reliably.

One example of this is hemoglobin in humans and the hemoglobin in legumes (leghemoglobin), which are distant relatives from the same protein superfamily. Both serve the same purpose of transporting oxygen in the organism. Although both of these proteins have completely different amino acid sequences, their protein structures are virtually identical, which reflects their near identical purposes and shared ancestor.

Other techniques for predicting protein structure include protein threading and *de novo* (from scratch) physics-based modeling.

Another aspect of structural bioinformatics include the use of protein structures for Virtual Screening models such as Quantitative Structure-Activity Relationship models and proteochemometric models (PCM). Furthermore, a protein's crystal structure can be used in simulation of for example ligand-binding studies and *in silico* mutagenesis studies.

Network and Systems Biology

Network analysis seeks to understand the relationships within biological networks such as metabolic or protein–protein interaction networks. Although biological networks can be constructed from a single type of molecule or entity (such as genes), network biology often attempts to integrate many different data types, such as proteins, small molecules, gene expression data, and others, which are all connected physically, functionally, or both.

Systems biology involves the use of computer simulations of cellular subsystems (such as the networks of metabolites and enzymes that comprise metabolism, signal transduction pathways and gene regulatory networks) to both analyze and visualize the complex connections of these cellular processes. Artificial life or virtual evolution attempts to understand evolutionary processes via the computer simulation of simple (artificial) life forms.

Molecular Interaction Networks

Tens of thousands of three-dimensional protein structures have been determined by X-ray crystallography and protein nuclear magnetic resonance spectroscopy (protein NMR) and a central question in structural bioinformatics is whether it is practical to predict possible protein–protein interactions only based on these 3D shapes, without performing protein–protein interaction experiments. A variety of methods have been developed to tackle the protein–protein docking problem, though it seems that there is still much work to be done in this field.

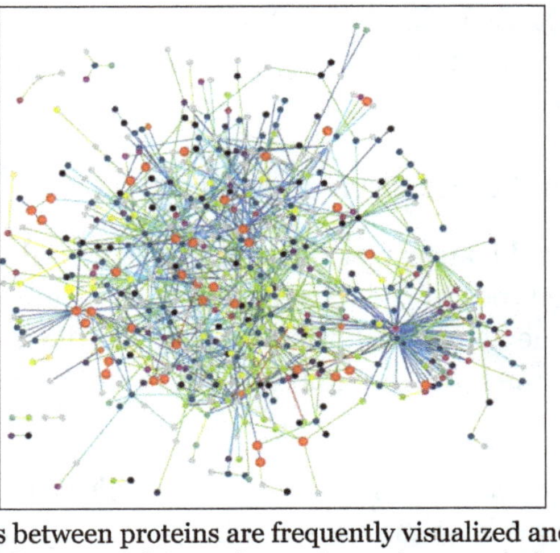

Interactions between proteins are frequently visualized and analyzed using networks. This network is made up of protein–protein interactions from *Treponema pallidum*, the causative agent of syphilis and other diseases.

Other interactions encountered in the field include Protein–ligand (including drug) and protein–peptide. Molecular dynamic simulation of movement of atoms about rotatable bonds is the fundamental principle behind computational algorithms, termed docking algorithms, for studying molecular interactions.

BIOMECHANICS

Biomechanics is the scientific discipline that applies the principles of mechanics to human movement. Biomechanics incorporates the concepts of engineering and physics in order to understand normal and pathological motion, physical performance, and mechanisms of injury. This is the study of how and why we move. By measuring the motion and forces applied to the body during human movement, potential causes of dysfunction can be identified. Typical Physical Therapy or Personal Training assessments do not utilize biomechanical assessments due to a lack of sufficient equipment and expertise to perform this type of assessment. A great deal of guesswork and assumption becomes necessary without the ability to measure force and motion during activity. Accurate assessment of walking, running, jumping, cutting or cycling activities are impossible without the aid of specialized equipment. The use of high-speed cameras, force measuring systems and EMG telemetry are required to accurately record and measure motion and forces.

BIOMEDICAL SIGNAL TRANSCEIVERS

A wireless transceiver provides a cost effective way to transmit biomedical signals to the various personal electronic devices, such as computers, cellular devices, and other mobile devices. Different kinds of biomedical signals can be processed and transmitted by these devices, including electroencephalograph (EEG), electrocardiograph (ECG), and electromyography (EMG). By utilizing wireless transmission, the user gains freedom to connect with fewer constraints to their personal devices to view and monitor their health condition.

Types of Biomedical Signal Transceivers

Different biomedical signal transceiver device types can be designed. There are several distinctions between the types of the devices and their operation. The distinctions can be based upon how the device is powered and how the device communicates. Despite these design differences, the hardware makeup of a biomedical signal transceiver is very standard.

Before going deeper into the details on the types of biomedical signal transceivers, it is important to understand how the device will operate. Typically, a biomedical signal transceiver device will have two main components, the transmitter and the receiver. The transmitter has several sub systems, including: signal acquisition, amplification, filtering, and as the name dictates, transmitter. The receiver subsystem will receive the signal from the transmitter, perform any required analysis on the signal, and then display the results. The transmitter is separate from the receiver, such that the transmitter can acquire the bioelectric signal and transmit to another device for remote viewing and analysis.

Existing biomedical transceivers are can be separated into two groups describing how they are powered; Radio Frequency (RF) and battery powered. In RF powered transceivers, an inductive link with external controller allows the transmission of power and commands. A common application of the RF powered transceiver is the transcutaneous neural recording arrays. In battery-powered transceivers, an onboard battery is utilized power source. This battery can be either disposable or rechargeable, depending on the device application. The use of a battery allows using higher frequencies for transmission and improved data rates can be achieved.

Another way to group biomedical transceivers is by their communication style. Biomedical transceivers can communicate either wirelessly or in the traditional wired connection. Not only can the device transmit the biomedical signal, but some devices have communication between the transmitter and receiver for not only biomedical information, but also any feedback or control signals. In this case, both subsystems are acting like transceivers.

Applications of Biomedical Signal Transceivers

Biomedical signal transceivers can be very useful in the monitoring devices and biotelemetry. There are several applications for these devices and their design is as unique as the application. These applications also utilize wireless communications to improve the system and the ease of use.

A health monitoring system which acquires and transmits the vital signals of a patient remotely to a hospital or medical professional can be very useful. This application of biotelemetry can allow for a patient to leave the hospital or clinic, but still have their health monitored remotely. Various bioelectric signals can be recorded from the patient's body and transmitted such as EEG, ECG, body temperature and blood pressure. Biomedical signal transceivers do not have to be limited to just an overall health monitoring device. These transceivers can also have more specific functions that can allow for more in depth analysis, depending on the application.

An ECG monitoring system is a great example of an application of biomedical signal transceiver. When the device is developed wirelessly, patients can monitor their heart signal via a mobile device, while having the electrodes and transmitter attached to their body. Furthermore, a warning system can be designed that can inform the patient about any abrupt abnormality in the heart. As with the health monitoring system, these heart anomalies can also be reported remotely to medical professions who can more appropriately analyze the patient's condition in real time. Another application of biomedical signal transceivers is to monitor the drug and medication usages in the patients remotely.

Analog Hardware Design

One of the most important parts of any biomedical signal transceiver is the analog hardware. Using this circuitry the biomedical signal is acquired, filtered, and amplified to an appropriate level. Along with this circuitry, the power for the system needs to be addressed. Finally, safety for both the system circuitry and the patient must be understood and taken care of during design to protect the device as well as the users.

Electrodes

For all biomedical devices that operate outside of the human body (i.e. non-implantable devices), electrodes are critical components. It is through these electrodes that the bioelectric potentials of the body are collected and transmitted to the measurement and analysis device. In this sense, electrodes are the initial part of any biomedical device.

Electrode Placement

The placement of the electrodes is dependent on the desired physiological signal. For example, to acquire an ECG signal, the electrodes may be placed in a triangle formation around the heart, creating the Einthoven's triangle. Each of the various bioelectric signals has a standard electrode placement on the body that must be understood and followed before acquiring signals. It is important to have the lead placement correct as noise and distortions will result from electrode misplacement. In some cases, only a small subset of the electrodes is required for signal acquisition. For example, in EEG recording the number of electrode may be from 1 to 128 electrodes. This is dependent on the application or use of the biometric signals by the device.

Electrode Make-up and Selection

An electrode is simply a mechanism that is used to make an electrical connection to a non-metallic surface. The electrodes have a common makeup, no matter what application or types of signals that are being acquired. Disposable electrodes have a very generic composition and purpose. Using an adhesive, the electrode is attached to the skin, which reduces the risk of noise artifact being introduced into by signal by electrode movement. Additionally, the electrode contains a gel that lowers the skins resistance and is therefore produces a better signal measurement. This allows for the metallic surface to conduct the signal onto the biomedical device.

There are several commercially available electrodes on the market today. The electrode performance will vary from company to company, and from part to part. It is essential to find an electrode that is appropriate for the application, all while keeping quality and price per electrode in mind.

The next step is to develop circuitry to prepare the analog signal for analysis. This will be accomplished by both amplifying and filtering the weak bioelectric signals. These steps are critical for all types of biometric signals.

Amplifier and Filter Design

When the bioelectric signals are acquired from the human body by the electrodes, the signals are very weak (small amplitudes). Because of their small amplitudes, these signals have little use to any biomedical sensor or system. However, if these signals are amplified to an appropriate level, they can be detected and read accordingly for analysis. The amount of amplification, termed gain, is determined by system specification and is dependent on the signal being measured, and other circuitry requirements. Another critical aspect of the signals that are acquired from the electrodes is the amount of noise in the signal. For proper signal analysis, these errors and noise need to be removed from the signal. The next sections go over the design of amplifiers and filters, all of which accumulates into the filter and amplifier circuitry design.

Amplification

To perform any sort of analysis on a bioelectric signal, the signal needs to be amplified to a level which an analog to digital converter (ADC) can sample the data with a high resolution. As well, the amplifier circuitry needs to include level shifting circuitry such that the signal is positive and has a similar dynamic range as the Analog to Digital conversion. The overall gain is defined by designer based upon the signal requirements.

The first stage in any amplification circuit is the instrumentation amplifier. This amplifier is a critical component for several different reasons, and has many applications outside of just biomedical devices. For one, the instrumentation amplifier acts as a buffer circuit by having large input impedance. This allows for very little current to be drawn from the source. By design, the instrumentation amplifier has a very large Common Mode Rejection Ratio (CMRR). The CMRR simply measures the tendency of the amplifier to reject a signal that is common between the two input pins. This is important for application in biomedical devices since the signal measurement is not coming from one electrode, but actually the difference between two electrodes (i.e. one lead). As well, the instrumentation amplifier will cancel out common noise between the two signals.

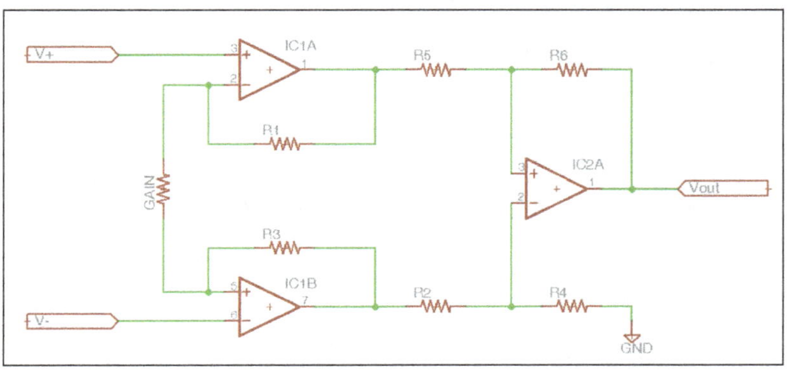

Instrumentation amplifier layout, utilizing three op-amps.

There are two ways to implement an instrumentation amplifier; a single IC package or a connection of several separate operational amplifiers (op-amps). Both are viable options, but the choice depends on the cost and efficiency of the choices. In most cases, it can be more efficient to use a single IC package instrumentation amplifier instead of the multiple op-amp design. Figure shows an example of cascaded op-amp design for the instrumentation amplifier.

The next stages of the amplifier circuitry are very simple in nature. The amplifier stages only purpose is to increase the amplitude of the bioelectric signals via amplification. Typically, one will employ only non-inverting amplifier configuration so that the signal does not be become inverted or out of phase. The proper number of amplifier stages typically is user defined, but can also be determined based upon the filter configuration. Both of these topics will be covered in the following sections.

The final stage of the amplifier circuitry is a level shifting circuit. The purpose of the level shifting circuit is to shift the negative components of the signal to a positive level. This also shifts up the positive voltage components of the signal as well. This circuit is critical as an analog to digital converter (ADC) on a microcontroller cannot read negative voltages. Thus, the signal would not be accurately converted, and the bioelectric information would be lost. There are several ways to implement this circuit; for example, one can use a non-inverting summing amplifier, which is illustrated in figure.

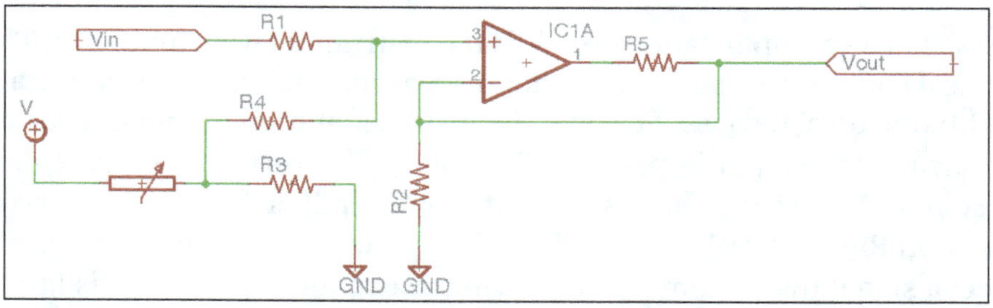

Summing amplifier design.

There are several other designs available, and they all require the use of an amplifier. The level shifting circuit in figure is a great example of this.

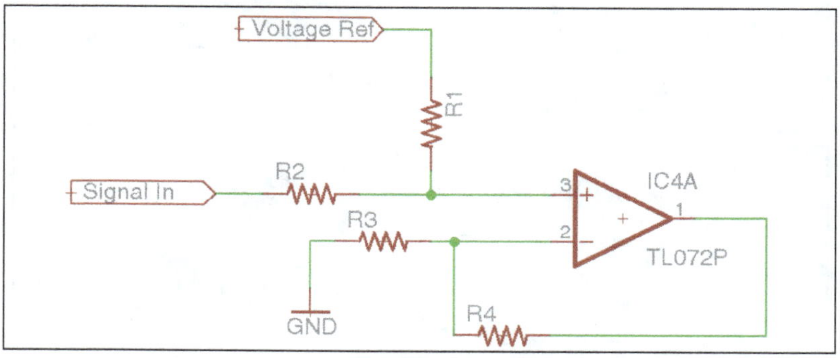

Level Shifting Circuit.

Both of these circuits will allow for the signal to be shifted to an adequate level. To do so, the resistor values will need to be designed such that the values that will allow for proper shifting. These values will also result in gain, if required. For example, if one does not want any gain from the level shifting circuit in figure (i.e. a gain of 1), simply follow the following guidelines:

- R1 = R4
- R2 = R3

If other values of gain (A) are required, the following equation should be considered:

- A = (R1/R3)x(R3+R4)/(R1+R2)
- R1 = R3
- R2 = R4
- A = (R4/R1)

This will allow to tune the circuit as required to shift the voltage. This level shifting circuit should be used if the exact value of shift is known; otherwise, the summing amplifier circuitry should be considered to allow for variance in the shift voltage. The potentiometer in summing amplifier allows the user to vary the voltage divider, which thus varies the shifting voltage level.

Amplifier Selection

There are many op-amps on the market for use in the amplification of bioelectric signals. However, many of the more common amplifiers, such as the common 741 op amp, do not produce ideal response, especially for bioelectric signals. This is due to the 741 design, or any other op amp that utilizes Bipolar Junction Transistors (BJT) in the first stages of the amplifier. Unlike MOSFET (or other FETs) the BJT will draw current from the signal, thus affecting the signal. As well, there are leakage currents from the BJT that will also hinder the signal. Thus, it can be advantageous to utilize an op amp that uses BiMOS technology. BiMOS is circuit design that use both BJTs and MOSFETs. Similar, BiCMOS can also be used, which simply is BJTs and CMOS. To determine the proper op amp to use in biomedical device one needs to look into the various specifications for the given op amp.

Filters

Filters are the other critical component of the analog hardware design for a biomedical device. By removing noise and artifacts from the signal, a precise and more accurate signal can be utilized by the signal analysis code. However, there are a lot of options and configurations for filters, and it can be tricky to determine what is necessary for the application at hand.

Before designing a filter, one needs to determine the frequency range of the bioelectric signals that are being measured. This is critical so that one can determine the required frequency response of the analog filters. Once this is determined, then the filters can be designed.

One of the most important filters, no matter what the frequency range of the bioelectric signal is, is the 60 Hz (or 50 Hz outside of North America) notch filter, also known as a band stop filter. This filter removes the noise that is produced from the common AC wall outlet. There are several ways to design a notch filter, with both passive and active designs. The effectiveness of the filter depends on the design. The passive filter designs will not be as exact, and the cut off frequency will vary over time (passive components will vary over time). This can and will affect the signal integrity over time. If there is a substantial enough drift, actually information will be attenuated with the 60 Hz being freely passed. Active filters, even with their power requirements, are by far the best option for most biomedical device application. One very effective and efficient design is to utilize Texas Instruments' Universal Active Filter, the UAF42, in a notch filter configuration. This design is laid out in the data sheet for the component, which explains the proper design for a 60 Hz notch filter with the chip and selected resistor values. There are several other active filter design and options that can be utilized to attenuate the 60 Hz noise from the signal.

The next filter that needs to be designed is the high pass and low pass filters. With these two filters in the circuit, it creates a band-pass filter (the band will be the frequency range of the bioelectric signal). As mentioned before, it is critical that this range of frequencies corresponds to the range of frequencies of the measured bioelectric signal. When designing the overall circuit, commonly the high pass filter is placed before the low pass filter.

High pass filter design is quite straightforward. Since a high pass filter will pass any frequency above the cut off frequency, the filter theoretically has an infinite frequency response. As such, if one was to design an active high pass filter, op amp utilized in the design may limit this response, as the op amp has a maximum frequency output. Therefore from a theoretical view, a passive filter will have the best response. In all practicality, this is not the case, but high pass active filters are still important to use, as they still can be effective. Depending on the bioelectric signal being measured, a simple RC passive filter can be sufficient. An example of a passive high pass filter is displayed in figure.

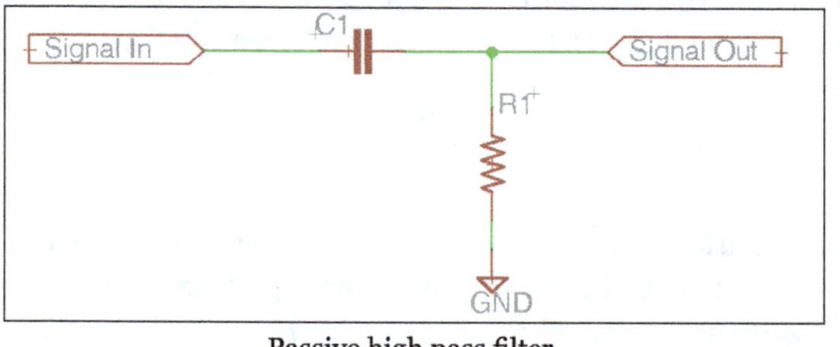

Passive high pass filter.

There are several possible common designs for the low pass filter. These include both active and passive filters. For these filters, there are several common types, including: Butterworth and Chebyshev. Each filter type has several configurations, with both active and passive designs. More commonly, for the low pass filter, an active configuration is utilized. As well, the Butterworth filter is typically used as it has an advantageous flat frequency response. There is also a choice of the filter order and configuration of the Butterworth Filter. A first order Butterworth low pass filter.

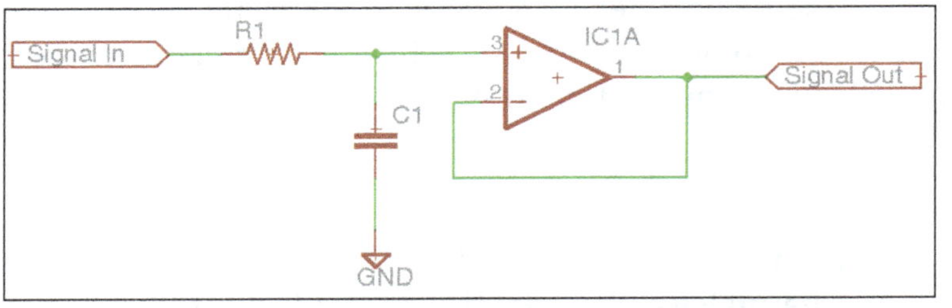

First Order Butterworth Low Pass Filter.

Typically, a filter with only a few orders will be utilized due to speed, cost, and space. A first order low pass Butterworth filter is perfectly acceptable choice for most applications. Another designer choice is the configuration of the filter. For active Butterworth filter, the Sallen-Key topology is an excellent choice. It allows for multiple orders, using only two op amps and several passive components. The design for the third order Butterworth filters is quite complex, and it involves calculating all values for the resistors and capacitors. There are several software packages that aide in the creation of these complex active filters.

Cascading Filters and Amplifiers

Now that both amplifiers and filters have been designed, the next step is the layout of the circuit in the most logical order. When designing circuitry that filters and amplifies a signal, there is a general rule of thumb to alternate stages of amplification and filtering. This is critical since one will introduce less noise into the circuitry as well as amplify less of the noise. This is why the design requires the cascading of op-amps and filters to alternate. Now, depending on the amount of gain that is required for the bioelectric signal, the following order can be used:

- Instrumentation Amplifier.
- High Pass Filter.
- Gain Stage.
- Low Pass Filter.
- Gain Stage.
- Notch Filter.
- Level Shift.

Naturally, the instrumentation amplifier will also have some gain. Most instrumentation op-amps have various levels of designable gain. To perform this type amplification, the de facto standard is to utilize a set of cascaded operational amplifiers. The need for cascaded stages will be explained later. When amplifiers are cascaded, one simply multiplies each of the gains together to determine the overall gain. Before designing the circuit, one needs to determine how much gain is actually necessary. The amount of gain will depend on what biometric signal is being measured, the ADC range, and other factors with that will vary from system to system.

Power

The voltage supply to the circuit components throughout the entire system is typically group together with the analog electronics design. Typically, there is a range of voltages are required through the system. For example, a microcontroller may require 5 V, a wireless transceiver may require 3.3 V, and the op-amps may require +/- 10 V. It is critical to design a system to effectively convert the input voltage to these different voltage values. It is also important to determine the total power that is required by the loads. There is a lot of DC – DC converters on the market, all of which have unique output power limits. In some cases it is perfectly acceptable to use voltage regulators instead of individual DC – DC converters.

Safety Issues

With any electrical device that is being interfaced with a human, safety is a critical part of the hardware design. Not only do you have to be concerned about the damage a human can do to the device (i.e. ESD) but also the harm the device can make to human. In the case of a wireless transmitting biomedical device, over voltage protection is not as important to the device as opposed to when the device is connected directly to the computer. This is because the wireless transmitter acts as an isolated buffer between the patient and the monitoring computer or device. This way, the highest voltage in the device will be the source, typically a low voltage battery. Even at the low voltages of a battery, some protection is necessary for the device. Typically, this is performed using diodes that are designed into the circuit to only conduct when there is an over-voltage event. Naturally, these diodes are placed near the inputs, such that it is the first/last components before passing to the patient. This way, when the voltage is above the forward break down voltage (0.7 V for a silicon diode) the diode will then conduct. Since bioelectric signals have such small amplitude, the diodes do not disturb their signal.

Digital Hardware Design

Microcontroller and Digital Hardware Design

The digital hardware system has three major parts: Microcontroller Unit (MCU), In System Programmer (ISP), and a Wireless Module (WM). The MCU includes a built in analog to digital converter (ADC). The ISP provides that capability to update the code on the MCU that is already a part of the board. Finally, the wireless module is involved in wireless data transmission. This communication is typically performed using Bluetooth. An illustration of this system is shown in figure.

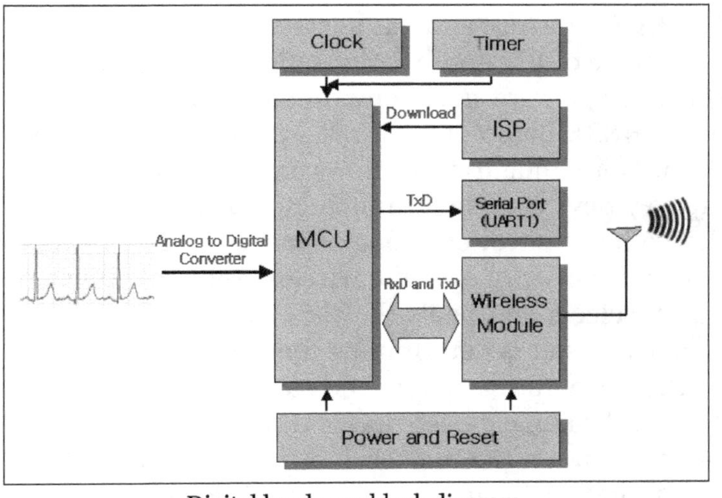

Digital hardware block diagram.

The MCU utilizes an 8-bit Reduced Instruction Set Computer (RISC), which has the advantages of simple commands, fast working speed, and low power consumption (2.7~5.5V). For example, the Atmel ATmega 128L has 16 Million Instructions per Second (MIPS) of performance. In addition, the ATmega MCU has 128 KB of In-System Programmable Flash with Read-While Writing capability. This is a type of flash memory that is 4 to 12 times faster than a general MCU. The ADC is responsible for converting continuous analog signals to digital signals. The ATmega MCU ADC has 8 channels and 10 bit resolution. This MCU also supports 16 different voltage input combinations and fast conversion time of 13~260us. Naturally, all MCUs are different, and these specifications will vary from MCU to MCU.

The ISP is the physical interface for programming the code on flash memory and EEPROM on the microcontroller. This hardware interface uses three signal lines: Master-Out-Slave-In (MOSI), Master-In-Slave-Out (MISO), and Clock (CLK). Once the reset pin on the microcontroller is set low, the code will updated via the ISP.

There are two possibilities to transmit the data from the microcontroller to a display device. One way to do this is via serial communication through the MAX232 IC. This IC will convert a TTL or CMOS signal into serial communication voltage level. This transmitting option requires a serial port and cable to transmit the data. In this sense, the communication is wired. Another option to transmit the data is via a wireless connection. This wireless communication is typically performed using a Bluetooth connection. This connection is created using a Bluetooth wireless module. In later sections, Bluetooth will be explored further. Other forms of wireless communication can be used, as per the system's requirements.

Wireless System Characteristics

Wireless system (Bluetooth) uses a 2.4GHz band for short distance and low power consumption communication. Bluetooth is used for its high reliability and low cost. It is supported by AT-Command and has a transfer rate of approximately 1 Mbps to 3 Mbps. Another feature of Bluetooth is its ability to guarantee stable wireless communication, even under severe noisy environment, by use of Frequency Hopping Spread Spectrum (FHSS). Bluetooth utilizes a packet based protocol with a master slave configuration. This configuration allows a wireless system to connect and communication to up to 7 devices. The terminology of master and slave is very straight forward. One device has control over one or more devices. In this application, the wireless transmitter has control over the wireless receiver. During communication, the transmitter first selects and pools the slave. This is termed the "inquiry" stage of communication, as master is determining the devices that it can connect to. Once the master pools the slave, the slave responds to the communication. Then the communication enters the "paging" phase that allows for the devices to synchronize the clock and frequency between the master and the slave. Once the master and slave modules are paired, the master module provides information to slave module with the master module's address. Once this is complete, the communication between devices can begin.

Table: Bluetooth 2.0 specifications.

Part	Specification
Bluetooth spec.	Bluetooth specification 2.0 Support
Communication distance	10M
Frequency Range	2.4 GHz ISM Band
Sensitivity	-83 dB m(Typical)
Input Power	3.3V
Current Consumption	48mA(Max)
Communication Speed	1,200bps~230.400bps
Antenna	Chip Antenna
Interface	UART(TTL Level)

To perform the Bluetooth communication, a Bluetooth Module is utilized. This transceiver allows for communication between microcontroller and display device. For example, a WINiZEN Bluetooth RS232 wireless module is used for the Bluetooth communication. The data transmission and receiving power consumption of this Bluetooth module is 18 to 30mA for transmission and 21~33mA for reception. Data transmission/reception is achieved for up to 10 meter distances. The WINiZEN Bluetooth module is operated at low power (3.3 V), and has the dimensions of 18x20x12 (mm). This size can be visualized in figure.

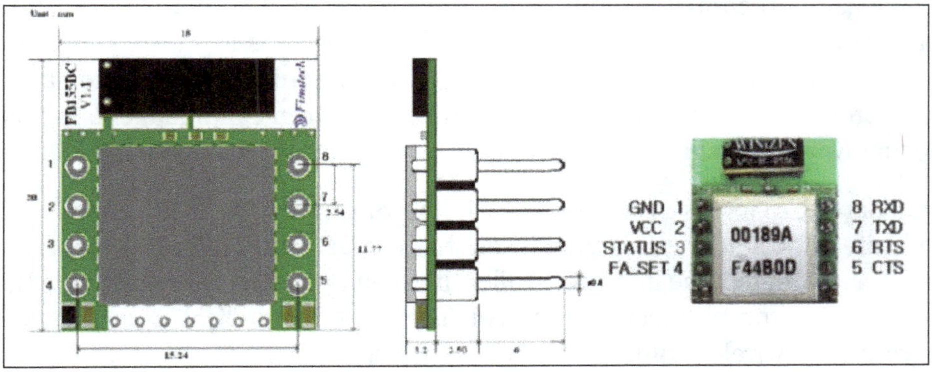

Bluetooth Module Dimensions and Pin Assignments.

The advantage of this, and any other small Bluetooth module, is the internal chip antenna that is used for short distance wireless communication. This is important since the transmitter or receiver device does not require an external antenna.

Firmware and Data Communication

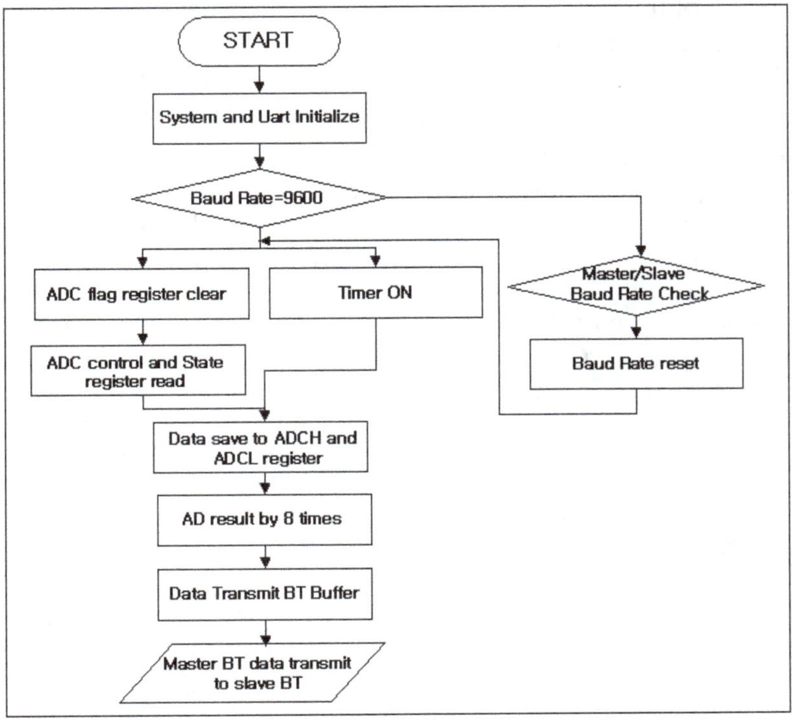

Firmware Block Diagram.

The development of the firmware for the microcontroller is based upon the microcontroller that is being utilized in the design. As well, the code can be written in a verity of languages, typically Assembly and C, which again is dependent on the microcontroller selected. To continue the example microcontroller that was used in previous sections, Atmel ATmega microcontroller code was created using AVR Stdio4.0 and AVR ISP. The code was written in the C-language. AVR Stdio4.0 which is a professional Integrated Development Environment (IDE) is used for writing, simulation, emulation and debugging. As a compiler, it also changes the firmware code from C-language to Hex code. An example firmware flow-charts for a biomedical signal transceiver is illustrated in figure.

The digital hardware works as follows. Once data from analog hardware circuitry has been amplified, filter, and shifted, the signal is sent to the ADC on the microcontroller. After the signal has been sampled, the microcontroller then sends the digital signal to the Bluetooth module via UART communication protocol. Then, the biomedical signal is transmitted wirelessly through wireless Bluetooth module, utilizing the master-slave configuration.

ECG Signal Processing: A Practical Approach

As an example of biomedical signal processing in a transceiver device ECG signal analysis for real-time heart monitoring will be discussed. Following the discussion on previous sections, this section particularly describes where real-time ECG monitoring using hand-held devices, such as, smart phones or custom-designed health monitoring system. ECG analysis is one of the very first areas of physiological measurement where computer processing was introduced successfully. ECG has been widely used as the diagnostic measurement for heart monitoring. Typically cardiologists used to interpret the ECG tracing for identifying abnormalities related to function of heart. With the advancement in biomedical signal processing, it is now reality that smart monitoring system can mimic the rules cardiologists applies in order to monitor heart health in real time. Signal Processing and machine learning techniques plays an important role in this purpose. Hence, this type of smart devices would be useful in monitoring patient's health in a more efficient way.

We have described a low-cost real-time biomedical signal transceiver/monitoring system which could also use ECG as one of the physiological parameter to monitor health.

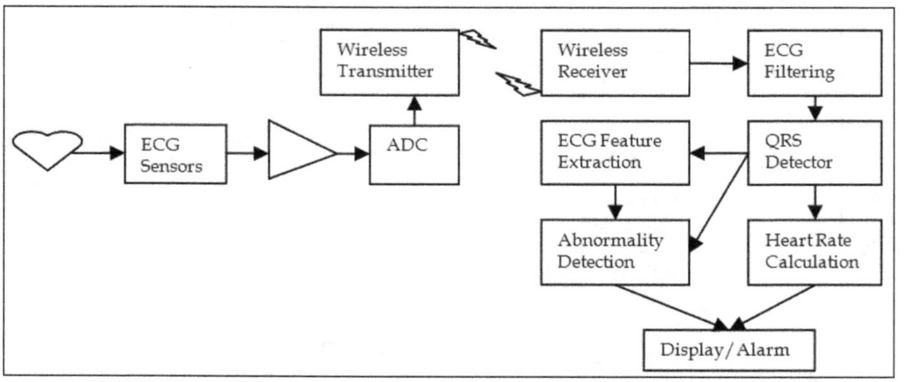

Block diagram of biomedical signal transceiver for real-time acquisition, processing and heart condition monitoring system. The right portion of the system can be implemented in smart phone or PDA for smart health monitoring.

In the following sub-sections, we will discuss the underlying ECG signal processing techniques used for identifying heart abnormalities. Typical steps are preprocessing of ECG, wave amplitude and duration detection, heart rate calculation, ECG feature extraction algorithms, and finally identifying the abnormalities. A typical block diagram is illustrated in following figure.

ECG Digital Filtering

ECG signal is usually corrupted with different types of noise. These are baseline wander introduced during data acquisition, power line interference and muscle noise or artifacts. Prior to applying different signal processing techniques for extraction different ECG features which are of clinical interests, data needs to be filtered in order to reduce noise and artifacts. Some of these filtering techniques with a practical approach where the methods can be easily implementable.

Power Line Interference

Appropriate shielding and safety consideration could be employed to reduce this particular type of noise in addition to analog filtering. After receiving signals at the receiver sides, it is preferred to get rid of this type of noise in the pre-processing step. Typically, band-stop (notch) filtering with cutoff, Fc=50/60 Hz would suppress such noise. Figure illustrates the magnitude and phase response of a digital second order Infinite Impulse Response (IIR) notch filter with cutoff frequency of 60 Hz.

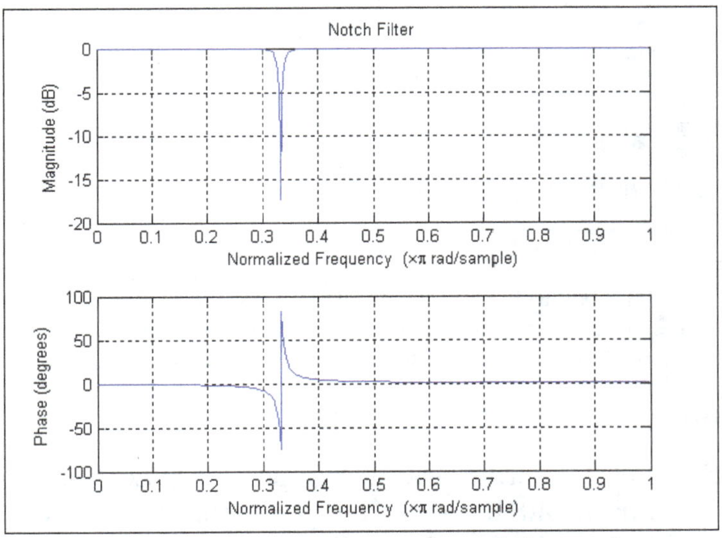

Magnitude and phase response of an Infinite Impulse Response (IIR) notch filter to remove 60 Hz power line noise.

Baseline Wanders Correction

Due to the error in data collection process, baseline as well as dc offsets could be introduced. Removal of baseline wander is required prior to further processing. There are several techniques that have been widely used such as filtering and polynomial curve fitting method. Since all data processing starts with visual inspection, as for offline ECG analysis, we should inspect the data for baseline wander and dc bias. DC bias can easily removed by subtracting the mean from the original signal. As for the baseline wander corrections, filtering method is most widely used.

The frequency of the baseline wander is usually below 0.5 Hz. This information particularly helps in design a high-pass filter in order to get rid of baseline wander. The design of a linear time-invariant high pass filter requires several considerations, most importantly, the choice of cut-off frequency and filter order. The ECG characteristic wave frequencies are higher than baseline wander. Therefore, a carefully designed high pass filter with cut-off frequency, Fc=0.5Fc=0.5Hz can effective remove the baseline. Baseline wander correction using a linear digital filter is shown in figure. Instead of a high-pass filter we used band-pass filtering. We applied a sixth order butterworth filter with cutoff 0.7-40 Hz. To avoid distortion, zero phase digital filtering was performed by processing the data in both forward and reverse direction. This can be done using MATLAB command filtfilt.

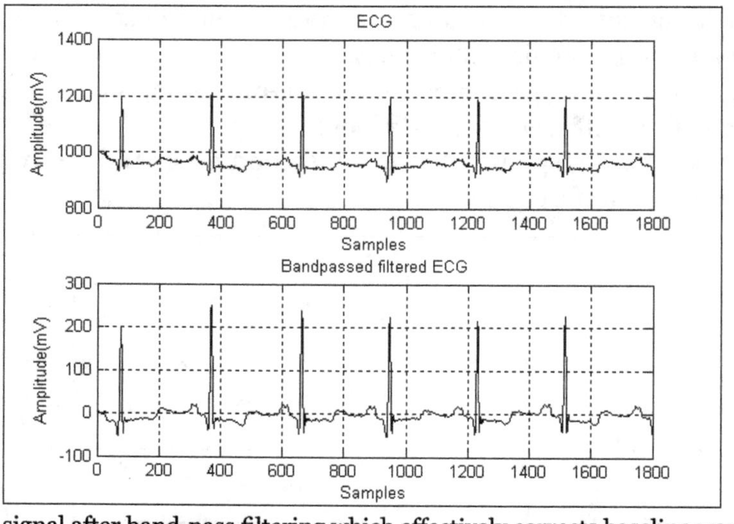

ECG signal after band-pass filtering which effectively corrects baseline wander and reduces low frequency noise and high frequency artifacts. Prior to applying bandpass filter the DC components were removed from ECGs.

QRS Detection

Detection of QRS complex is particularly most important in ECG signal processing. The information obtained from QRS detection, temporal information of each beat and QRS morphology information can be used for the improvement of performance for the other algorithms. Many methods have been developed for detection of QRS complex. Most of these are based on filtering, nonlinear transformation and decision rules as well as using template matching method. In recent approaches, wavelet transform and empirical mode decomposition based algorithms have been developed. We will discuss a widely used and robust real-time QRS detection algorithm popularly known as Pam-Tompkins algorithm.

The Pan-Tompkins algorithm consists of several steps. This is a single channel detection method in order to achieve better performance. In order to attenuate noise, the signal is first passed through a digital band-pass filter composed of cascaded high-pass and low-pass filters. The band-pass filtering corrects the baseline wander, reduces muscle noise, 60 Hz power line interference, T-wave interference. The cutoff frequency of this band-pass filter is an important design parameter. Another reason of this filtering is to maximize the energy of QRS complex. Pan and Tompkins suggested choosing 5-15 Hz as desirable pass band. The next steps involve nonlinear transformation in order to highlight the QRS complex from baseline. These steps are differentiation, squaring followed by moving window integration. Derivative stage provides the slope information of QRS complex. Squaring operation intensifies the slope of the frequency response curve. It also helps to restrict the false positives caused by unusually high amplitude T waves. The signal produced by the moving window integration operation provides the slope and width of the QRS complex. The choice of window sample size is an important parameter. Generally, it should be chosen such a way that the window size is same as the QRS width. The window width can be determined empirically making sure that QRS complex and T waves do not merge together. For practical application, we choose window of 30 samples (150 ms, sampling frequency was 360 samples/s). Finally, an adaptive threshold will be applied to identify the location of QRS complexes.

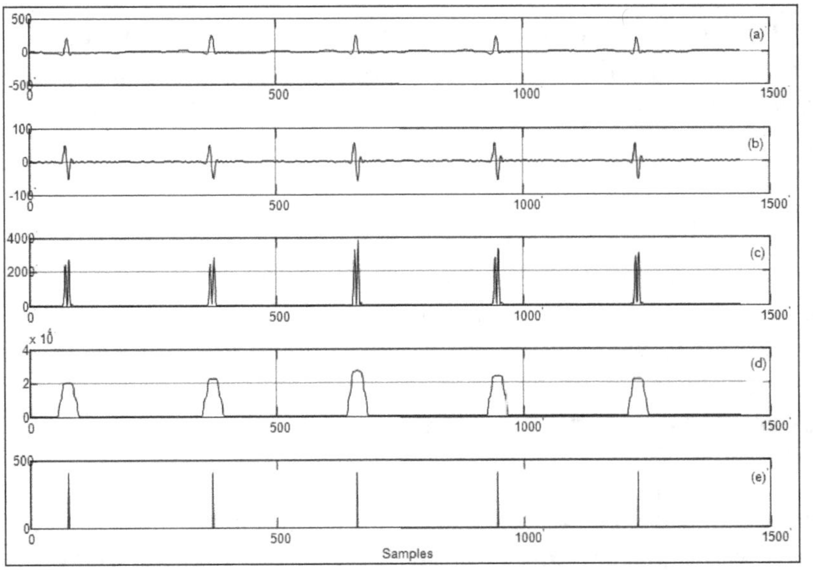

Typical steps of Pan-Tompkins algorithm for detecting QRS complex:
(a) band-pass filtered ECG signals; (b) after differentiation; (c) after performing squaring operation; (d) moving window integration; and (d) R peak detection.

The fiducial points of the QRS complex can be detected from the integration waveform. The time duration of the rising edge is equal to the QRS width. The maximum slope is the location of R wave peak. Q and S points are the start and end points of the rising slope respectively. The R peak detection is illustrated in figure.

Heart Rate Computation and Display

Heart rate is a vital sign to determine the patient's condition or well being. It can be easily computed from ECG. Heart rate can be computed using various methods, for example, over one minute time period simply by counting the number of heart beats would give us the heart rate in beats per minute. However, if the QRS or R wave peak detector misses some beats that would adversely affect the heart rate and give wrong counts. The heart rate monitoring tool should avoid such wrong results. Therefore, it is more appropriate to measure the time duration between two successive R peaks known as R-R intervals and computing instantaneous heart rate directly from R-R interval. In clinical settings, heart rate is measured in beats per minute (bpm). So the formula for determining heart rate from RR interval is as following:

$$\text{Heart rate (bpm)} = \frac{60{,}000}{\text{RR interval (ms)}}$$

Once the program computes the heart rate it can be presented on the display of the monitoring device. We have discussed a low cost biomedical signal transceiver, where ECG signal is one parameter for health monitoring. Using this wireless transceiver ECG signal can be transmitted to a hand-held device, such as, smart phone or PDAs. The smart phone will have applications capable of computing instantaneous heart rate or average heart rate of a pre-determined period of time and shown as a digital output in the display of the smart phone. The application would also be able to identify primary abnormalities such as premature atrial contraction (APC or PAC) or premature ventricular contraction (PVC) and present an auditory or visual alarm to the patient.

P and T Wave Detection

Once QRS locations are identified, two other fudicial marks in ECG, P and T wave peak can be detected using the R peak location information. For detecting T wave onset, peak and offset, we should define a search window in forward direction. The size of the search window can be function of heart rate calculated. For detecting P wave, a backward search is required. Search window size can be determined similarly.

Proposed an automatic method of wave boundaries detection in multilead ECG. The algorithm includes a QRS detector for multilead ECG, P and T wave detection. This method is also based on the QRS detection method of Pan and Tompkins. In preprocessing steps, a second order linear bandpass filter (cutoff frequency of 0.8-18 Hz, -3dB) is applied to the signal to remove baseline wander and attenuate high frequency noise. This signal is called ECGPB. Then a low-pass differentiator is applied to obtain the information about changes in slope. This differentiated signal is called ECGDER. Then moving-window integration will be performed. In this case, Laguna et al. had chosen the window size of 95 ms. Since P and T waves usually have lower frequency components than the QRS complex, a digital low-pass filter (cutoff frequency of 12 Hz, -3dB) is applied to the ECGDER signal. This filtering reduces the remaining noise and highlights the P and T wave components and the signal is called (DERFI). Now a search window of 150 ms is defined which starts 225 ms before the R wave position. In this search window, the algorithm searches for maximum and minimum values. The P wave peak is assumed to be occurred at the zero-crossing of the maximum and minimum values.

Similarly for T wave detection, a search window is defined which is a function of heart rate already calculated. In this search window, the algorithm searches for maximum and the minimum values. However, the T wave peak is assumed to be occurred at the zero-crossing of the maximum and minimum values. The T wave morphology varies hugely. The algorithm can also identify the type of T wave (for example, regular or inverted) by examining the relative position and values of the maximum and minimum values in the search window.

BIOMEDICAL SENSOR, DEVICE AND MEASUREMENT SYSTEMS

Sensors are very critical components in all devices and measurement systems. They have been widely used in a lot of fields such as science, medicine, automated manufacturing, environmental monitoring, and so on. Some cheap sensors are finding their ways applying into all sorts of consumer products, from children's toys, dishwashers to automobiles. To some extent, sensors are multidisciplinary and interdisciplinary field of endeavor.

There are a lot of terms which are often used for sensors including transducer, meter, detector, and gage. Defining the term sensor is a very difficult task. At present, there is not a uniform definition which is agreed by all of us. The most widely used definition is that which has been applied to electrical transducer by the Instrument Society of America: "Transducer —A device which provides a usable output is in response to a specified measurand." Furthermore, national standard of China points out that sensors consist of sensing component, converting device and electronic circuit. A transducer is more generally defined as a device which converts energy from one form to another. Output of sensor can be an optical, electrical, chemical, or mechanical signal. In the field of electrical engineering, the measurand is physical, chemical, or biological property or condition measured; hence output of biological signal should be an electrical signal, too.

The words sensor and transducer are both commonly used in the context of measurement systems, and often in an interchangeable manner. Transducer is used more in the United States, and sensor has great popularity in Europe and China. The blurring of lines between the exact meaning of sensors and transducers leads to a degree of confusion. Most but not all sensors are transducers, employing one or more transduction mechanisms to produce an electrical output signal. According to the basic sensing principle, sensors are classified into mechanical sensors, electrochemical sensors, biosensors, optical sensors, semiconductor sensors, magnetic sensors, and thermal sensors. From different viewpoints, there are different classifying methods. According to the physical parameters measured by sensors, sensors are classified into resistance displacement sensor, inductive displacement sensor, capacitive displacement sensor, piezoelectric pressure sensor, laser interferometer displacement sensor, bore gagging displacement sensor, ultrasonic displacement sensor, optical encoder displacement sensor, optical fiber displacement sensor, optical beam deformation sensor, flow sensor, imaging sensor, temperature sensor, intelligent sensor and chemical ingredient sensor. Biomedical sensors are used to gain the information on body and pathology, which is a branch of biomedical engineering. Biomedical sensors are classified into physical sensor, chemical sensor and biosensor. Physical sensor could be employed to measure blood pressure, body temperature, blood flux, blood viscosity, biological magnetic field, etc. Chemical sensor is utilized to detect the ingredient and concentration of body liquid such as PH value, Ca^+ concentration, glucose concentration, etc. Biosensor is used to sense enzyme, antigen, antibody, hormone, DNA, RNA and microbe. In nature, biosensor is a kind of chemical sensor, which is mainly used to detect biological signals.

Sensor Package and Specifications

Packaging of certain biomedical sensor is an important consideration during the design, fabrication, and use of the device. Obviously, the biomedical sensor has to be safe, soft, and reliable for biomedical sensors often touch the body skin or inner organs of patients. In the development of implantable biosensor, an additional key issue is to consider the biocompatibility of sensor and operational lifetime in body. When a biomedical sensor is implanted into the body, it inevitably contacts with body fluids. Then body will affect the function of biomedical sensor, or sensor will affect the site that it is implanted. For example, protein absorption and cellular deposition can alter the permeability of sensor packaging that is designed to both protect sensor and allow free chemical diffusion of certain analytics between body fluids and the biosensor. Unsuitable packaging of implantable biomedical sensor could lead to drift and a gradual loss of sensor sensitivity and stability overtime. Furthermore, inflammation of tissue, infection, or clotting in a vascular site could produce some harmful or adverse effects on biomedical sensor. Hence, the material used in the construction of sensor's outer body must be biocompatible because they play a crucial pole in determining an overall performance and longevity of implantable biomedical sensor. One convenient method is to utilize various polymer covering material and barrier layers to prevent the toxic sensor components from coming into body. It's very important that packaging material of biomedical sensor must prevent the chemical diffusion of harmful ingredient between biomedical sensor and outer body.

Accurate medical diagnostic procedures require the stringent specifications on the design and use of biomedical sensor. Depending on the intended applications, the performance specifications of biomedical sensor may be evaluated to ensure that the measurement meets the design specifications.

In order to understand sensor's performance characteristics, it is very important to learn some of the common terminology associated with sensor specifications. The following definitions are commonly used to describe sensor characteristics and select sensor for particular applications.

Measurement Range

The range of sensor corresponds to the minimum and maximum operation limits that sensor is expected to measure accurately. For example, a pressure sensor may have a nominal performance over the operating range from 0 Pa to 10MPa.

Sensitivity

Sensitivity refers to the ratio of output change for a given input change. Another way to define sensitivity is to find the slope of calibration line relating the input to the output, as illustrated in figure. A high sensitivity implies that a small change in input causes a large change in output.

For example, a pressure sensor may have a sensitivity of 0.4V/Pa0.4V/Pa; that's to say, the output of this sensor will change 0.4V for 1Pa change in input pressure. If the calibration curve is linear seen in figure (a), then sensitivity of sensor will be constant, whereas the sensitivity of sensor will vary with the input when the calibration is non-linear, as in figure (b). Alternatively, sensitivity can be defined as the smallest change in input that will result in a detectable change in sensor output.

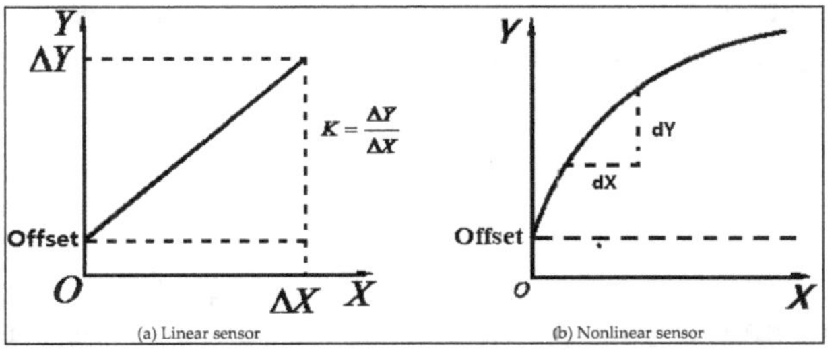

Input versus output calibration curve of a typical sensor.

Accuracy

Accuracy refers to the difference between the true value and the actual value measured by sensor. Classically, accuracy is expressed as a ratio between the preceding difference and the true value; it is specified as a percent of full-scale readings.

Precision

Precision refers to the degree of measurement reproducibility under the same conditions.

Very reproducible readings indicate a high precision. Precision should not be confused with accuracy. For an example, measurement may be very precise but not necessary accurate.

Resolution

When the input is increased from some arbitrary nonzero value, the output of a sensor will not change until a certain input increment is exceeded. Accordingly, resolution is defined as the smallest distinguishable input change that can be detected with certainty.

Reproducibility

Reproducibility describes how close measurements are when same input is repeatedly exerted into same sensor under same conditions. When the range of measurement is small, the reproducibility is very high. For example, a temperature sensor may have a reproducibility of±0.1V/°C for a measurement range from 20 °C to 80 °C. Here, what need to be noticed is that reproducibility can vary depending on the measurement range. In other words, readings can be highly reproducible over one range and less reproducible over a different operating range.

Offset

Offset refers to the output value when input value is zero, seen in figure (a) and (b).

Linearity

Linearity of sensor also called nonlinear error of sensor's characteristic curve; it is a measurement of the maximum deviation between calibration curve and fitting curve. Usually, linearity of sensor is expressed as a percent of full-scale readings or a percent of the actual readings. Linearity could be expressed as the following equation:

$$\sigma_{Linearity} = \pm\frac{\Delta L_{max}}{Y_{F.S}} \times 100\%$$

Here, $\sigma_{Linearity}$ —linearity of sensor; $\Delta L_{max}=\max(V_{cal}-V_{fit})$, ΔL_{max} is the maximum error between calibration line and fitting line; $Y_{F.S}$ is the full-scale meaning value of sensor, $Y_{F.S}=Y_{max}-Y_o$, Y_{max} is the maximum deviation of output, Y_o is the deviation without any input value.

Response Time

The response time indicates that the time it takes a sensor to reach a percent of its final steady-state value when input of sensor is changed. For example, it takes 10 seconds for pressure sensor to reach 95 percent of its maximum value when a change in pressure of 1Pa is measured. Ideally, a short response time indicates the ability of a sensor to respond quickly to change in input.

Drift

Drift refers to the change in sensor reading when the input keeps constant. Drift is divided into temperature drift and zero point drift. Zero point drift refers to the output without any input or with a constant input. Zero point drift could be expressed as the following equation:

$$D_{zero} = \frac{\Delta Y_o}{Y_{F.S}} \times 100\%$$

Temperature drift refers to the change of output with the change of temperature. It means the deviation of sensor output, which could be expressed as the following equation:

$$D_{Temp} = \frac{\Delta Y_{max}}{Y_{F.S} \times \Delta T} \times 100\%$$

Here, ΔT is the change of temperature.

Hysteresis

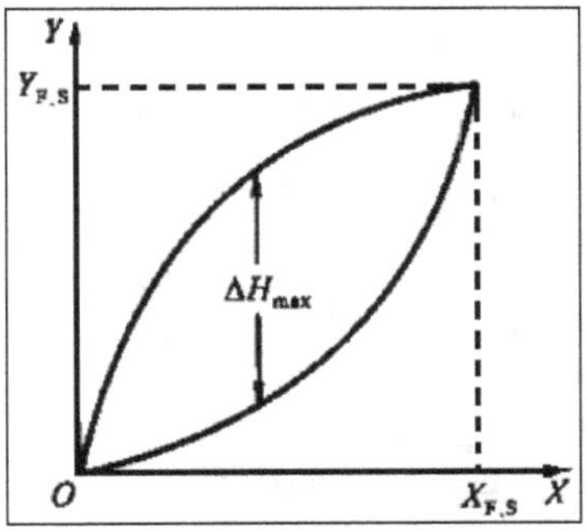

Input versus output response of a sensor with hysteresis.

In some sensors, the input-output characteristic follows a different nonlinear trend, depending on whether input increase or decrease, as in figure. For example, a certain pressure sensor could produce a different output voltage when the input pressure varies from zero to full scale and then back to zero. When the measurement is not perfectly reversible, the sensor will show its hysteresis. If a sensor exhibits hysteresis, the input-output relationship is not unique, but depends on the direction change to the input value of sensor.

Special Features of Biosensor

Biosensor is a kind of device which senses biomaterial and its concentration, and which converts the biosignal into electrical signal. Biosensor has the function of acceptor and converter, which configuration is seen in figure. In biosensor, the physicochemical change of the biologically active material resulting from the interaction with the analyte must be converted into an electrical output signal by an appropriate converter.

Biosensor's sensing components mainly have enzymes, cells, antibodies, DNA (Deoxyribonucleic acid), chemical electrode, microbe and other biologically active agents in analytical devices. In the course of detecting the parameters of analytes, biomaterial should be always immobile. In order to develop biosensor, some biotechnology has to be studied and applied, such as DNA biosensor, PH sensor, microelectrode, and so on.

The special features of biosensor are the following:

- Biological active material immobilized is used as catalyst, and expensive reagents could be repeatedly used to detect same biological parameters.
- Biosensor has intensive specificity. Biomaterial only senses definitive ingredient and it is not affected by color and concentration of measured material.
- Biosensor could quickly analyze the result of the measurand.
- Biosensor's accuracy is very high, which relative error could reach one percent.
- Biosensor's analyzing system is very simple.
- The cost of biosensor is very low.

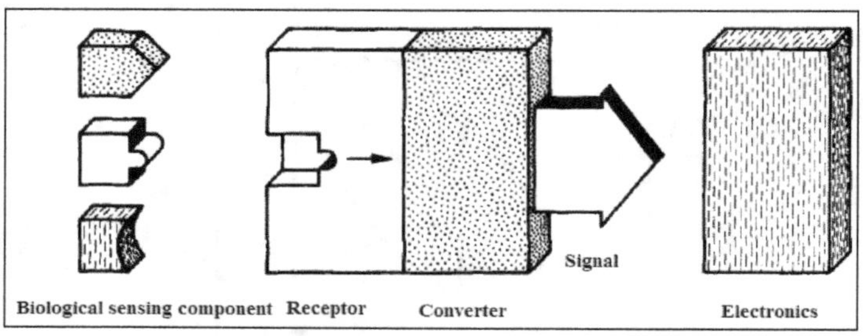

Common configure of a biosensor.

According to biological sensing component, biosensor may be divided into five classes: enzyme sensor, microbe sensor, cell sensor, tissue sensor, and immune sensors. According to the signal converter of biosensor, biosensor may be also divided into five classes: bioelectrode sensor, semiconductor biosensor, optical biosensor, piezoelectric biosensor and thermal biosensor. According to the interaction between sensing component and measured material, biosensor can be divided into two classes: affinity biosensor and catalytic biosensor.

Biomedical Sensor's Application

In biomedical field, main applications of biomedical sensor are as follows:

- Detecting the information of clinical chemistry. In the field of medical clinic and basic research, the biology's information needs to be detected to ensure the present state of given biology. For example, before operating on a patient, a doctor needs to know the body temperature and blood pressure. Under this condition, clinic thermometer and blood sensor has to be employed to help doctor quickly detect body temperature and blood pressure of patient.

- Continuously monitoring some parameters of biology outside and inside. In biomedical field, heart frequency has to be monitored continuously by heart sound sensor for a few days after operation. In military, some viruses need to be found by biosensor to hold back the attacking from enemy.
- Control: In medicine, people usually utilize some parameter detected by biomedical sensor to control or adjust physiological course of body. In the food industry, biomedical sensor could be utilized to measure some enzyme and its concentration to control the process of fabricating food and to analyze the nutritional ingredient of food. In military, biomedical sensor could be employed to detect the situation of battle field to adjust the strategy of spying or attacking enemy.

Of course, biomedical sensor such as PH sensor could be also employed to detect our atmosphere and condition to improve our living situation.

Biomedical Sensors

Biomedical Sensor Classification

Many different kinds of sensors can be used in biomedical application. According to the sensing principle in biomedical application, biomedical sensors can be classified into physical sensors and chemical sensors.

It's possible to categorize all sensors as being physical or chemical. In the case of physical sensors, quantities such as geometric, mechanical, thermal, and hydraulic variables are measured. In biomedical applications these variables can include things such as muscle displacement, blood pressure, core body temperature, blood flow, cerebrospinal fluid pressure, and bone growth velocity. Two types of physical sensors deserve special mention with regard to their biomedical application: sensors of electrical phenomena in the body, usually known as electrodes, play a special role as a result of their diagnostic therapeutic applications. The most familiar of these are sensors used to pick up the electrocardiogram, an electrical signal produced by the heart. The other type of physical sensor that finds many applications in biology and medicine is optical sensor. These sensors can utilize light to collect information, and, in the case of fiber optic sensors, light is the signal transmission medium as well.

The second major classification of sensing device is chemical sensors. In this case sensors are concerned with the chemical quantities such as identifying the presence of chemical composite, detecting the concentration of various chemical species, and monitoring the chemical activities in the body for diagnostic and therapeutic application. A wide variety of chemical sensors are classified in many ways. Chemical sensors are used to detect chemical components being measured and chemical composition measured in the gas phase. Electrochemical sensors are utilized to measure chemical concentration, or more precisely, activities based on chemical reactions that interact with electrical systems. Photometric chemical sensors are optical devices that detect chemical concentrations based on changes in light transmission, reflection or color. Other types of physical chemical sensors such as the mass spectrometer utilize various physical methods to detect and quantify chemicals associated with biologic systems.

Table: Classifications of biomedical sensor.

Class of sensor	Biomedical sensor
Physical sensors	
	Geometric
	Mechanical
	Thermal
	Hydraulic
	Electric
	Optical
Chemical sensors	
	Gas
	Electrochemical
	Photometric
	Other physical chemical methods
Biopotential electrodes	
	Body surface biopotential electrode
	Metal plate
	Intracavitary and intratissue electrode
	Microelectrode
Bioanalytic (or biosensor)	
	Enzyme
	Protein
	Antigen
	Antibody
	Ligand
	Cell
	DNA

Although bioanalytic sensors are essentially chemical sensors, they are often classified as a separate major sensor category. These devices incorporate biologic recognition reaction such as enzyme-substrate to identify complex biochemical molecules. The use of biologic reactions gives bioanalytic sensors high sensitivity and specificity in identifying and quantifying biochemical substances.

Oxygen and Carbon Dioxide Sensor for Blood

Measurements of arterial blood gas (pO_2 and pCO_2) and pH are frequently performed by on critical patients in both the operating rooms and intensive care unit. They are selected and used by the physician to adjust mechanical ventilation or administer pharmacological agents. Such measurement could provide information about the respiratory and metabolic imbalance in the body and reflect the change of blood oxygen increment and carbon dioxide(CO_2) elimination.

Noninvasive sensors for measuring O_2 and CO_2 in arterial blood are based on the discovery that gases such as O_2 and CO_2 can easily diffuse from body skin. Diffusion occurs due to a partial pressure difference between the blood in the superficial layers of the skin and the outermost surface of the skin. Such idea has been used to develop two types of noninvasive electrochemical sensors pO_2 and pCO_2. The discovery that blood changes its color depending on the percent of oxygen has led to the development of several optical methods to measure the oxygen saturation in blood.

Oxygen Sensor for Blood

The method for measuring blood oxygenation is very great important in assessing the circulatory and respiratory condition of a patient. The blood from the lungs to the tissues in two distinct states transports oxygen. Under normal physiological conditions, approximately 2% of the total amount of oxygen carried by the blood is dissolved in the plasma. This amount is proportional to the blood pO_2. The 98% remain is carried inside the erythrocytes in a loose reversible chemical combination with hemoglobin (Hb) as oxyhemoglobin (HbO_2). Thus, there are two methods for measuring the blood oxygenation: either using polarographic pO_2 sensor or measuring oxygen saturation (the relative amount of hemoglobin dioxide HbO_2 in the blood) by means of an optical oximeter.

A pO_2 sensor, also widely known as a Clark electrode, is used to measure the partial pressure of O_2 gas in a sample of air or blood. This sensor is categorized as an amperometric sensor and requires an external polarization bias source. The measurement is based on the principle of polarography as illustrated in figure. The electrode utilizes the ability of oxygen O_2 molecules to react chemically with H_2O in the presence of electrons to produce hydroxyl (OH^-) ions. This electrochemical reaction, called an oxidation/reduction or redox reaction, generates a small current and requires an externally applied constant polarizing voltage source of about 0.6V.

Oxygen is reduced (consumed) at the surface of a noble metal (such as platinum or gold) cathode (this electrode is connected to the negative side of voltage source) according to the following the chemical reaction:

$$O_2 + 2H_2O + 4e \leftrightarrow 4OH^-$$

In this reduction reaction, one O_2 molecule takes four electrons and reacts with two water molecules, generating four hydroxyl ions. The resulting OH^- ions migrate and react with a reference Ag/AgCl anode (this electrode is connected to the positive side of voltage source), causing a two-step oxidation reaction as follows:

$$Ag \leftrightarrow Ag^+ + e$$

$$Ag^+ + Cl^- \leftrightarrow AgCl \downarrow$$

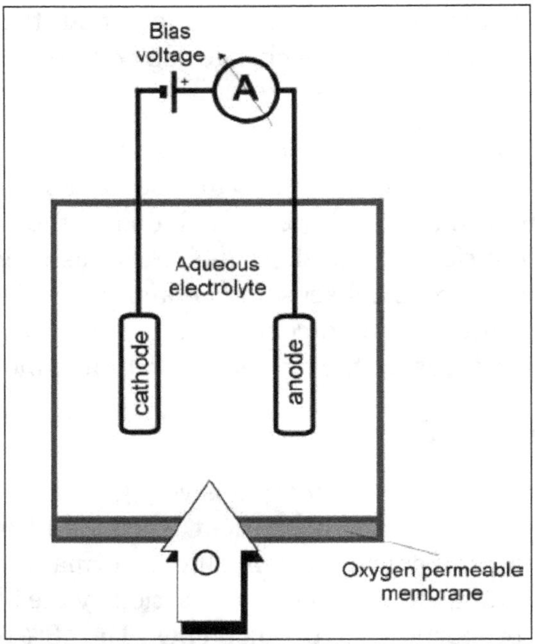

Sensing principle of Clark-type pO_2 sensor.

In this oxidation reaction, silver from the anode electrode is firstly oxidized to silver ions, and electrons are liberated to the anode. These silver ions are immediately combined with chloride ions to form a kind of compound precipitant silver chloride AgCl which precipitates on the surface of anode. The transient current between the anode and the cathode in the external circuit produced by this reaction is directly proportional to the number of O_2 molecules constantly reduced on the surface of the cathode. The electrodes in the polarographic cell are immersed in an electrolyte solution of potassium chloride and surrounded by an O_2-permeable or polypropylene membrane that permits gases to diffuse slowly into electrode. Thus, by measuring the change in current between the cathode and the anode, the amount of oxygen that is dissolved in the solution can be determined.

Carbon Dioxide Sensor for Blood

Electrodes for measuring partial pressure of carbon dioxide CO_2 in blood are based on measuring the pH as illustrated in figure. The measurement is based on the observation that it forms a weakly dissociated carbonic acid (H_2CO_3) that subsequently forms free hydrogen and bicarbonate ions when CO_2 is dissolved in water according to the following reaction:

$$CO_2 + H_2O \leftrightarrow H_2CO_3 \leftrightarrow H^+ + HCO_3^-$$

As a result of this chemical reaction, the pH of the solution is changed. This change generates a potential between the glass pH and a reference electrode that is proportional to the negative logarithm of the concentration of the carbon dioxide pCO_2 in the plasma.

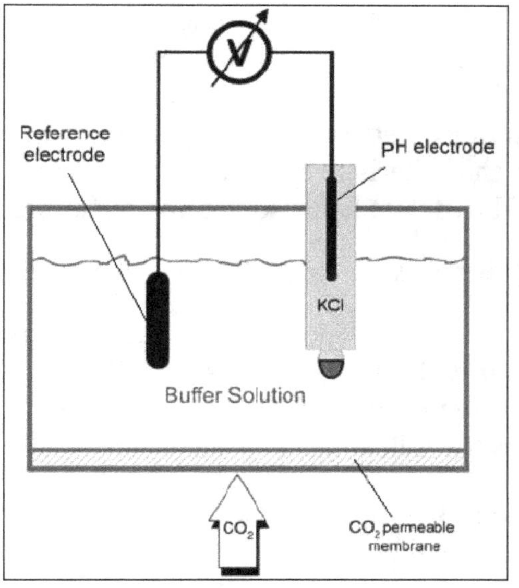

Sensing principle of a pCO_2 electrode.

Heart Sound Sensor

The expansion and shrinkage of heart necessarily lead to the vibration of artery that is formed by blood turbulence in vein. When the vibration of artery is transported to the surface of thoracic cavity, heart sound will take place. Heart sound is very valuable for doctor to diagnose many kinds of diseases in our body.

The range of heart sound is from 20Hz to 200Hz. Low limit frequency of heart sound could reach about 4Hz and high frequency limit is greater than 1000Hz. There are many kinds of medical heart sound sensors that are divided into two classifications: air conduction heart sound sensor and direct conduction heart sound sensor. Air conduction heart sound sensor consists of air chamber and common sensor. Such sensor has obvious defects: low sensitivity and easy disturb by surrounding circumstance. Hence, in clinic application, the most sensors applied are direct conduction heart sound sensor.

Piezoelectric Heart Sound Sensor

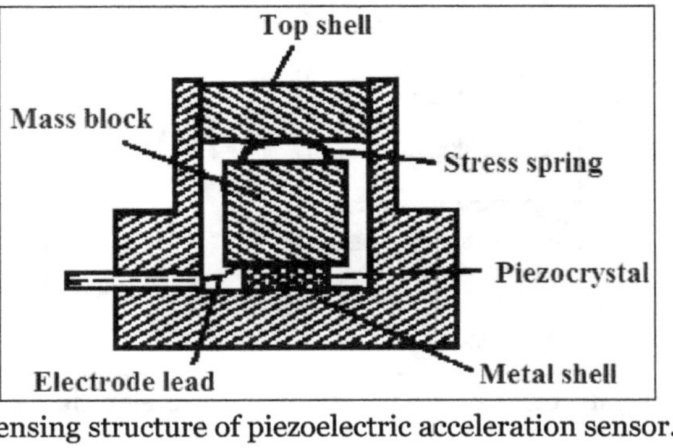

Sensing structure of piezoelectric acceleration sensor.

The sensing structure of piezoelectric acceleration sensor is illustrated in figure. Such sensor is used to measure heart sound. Its structure is very simple, which consists of vibration mass block and piezoelectric crystal. A stress spring is utilized to exert a certain stress on vibration mass block between top shell and mass block. Such method could timely adjust the linear characteristic of sensing component. This sensor's gravity is less than 30g, and is used to detect heart sound and buffeting from body organisms.

Fetus Heart Sound Sensor

Detecting fetus heart sound is very important in clinic application for doctor sometimes needs to grasp the present body status of fetus. PVDF piezoelectric thin film sensor is utilized to be fit for the measurement of fetus's heart sound as illustrated in figure. Its piezoelectric coefficient is the following:

$$d_{31} = (15 \sim 30) \times 10^{-12}\ C/N$$

$$d_{33} = -(5 \sim 8) \times 10^{-12}\ C/N$$

In this structure, silicon rubber converts the vertical motion of itself into the radial motion of PVDF piezoelectric thin film and then corresponding dynamic charge produced by PVDF thin film is proportional to the externally transient force. The voltage along thickness direction is output. Obviously, its work mode is d_{31} work mode. For both thin film and silicon rubber are very soft, they could well touch the skin of body belly. Then fetus heart sound is gained to judge fetus heart. Design requirement of PVDF heart sound sensor is as follows:

- In the same piezoelectric thin film, one part is used as driving electrode; another is utilized as receiving the sound wave from heart sound.
- Dynamic response characteristics on the surface of belly should be considered to well mate belly skin and sensor.

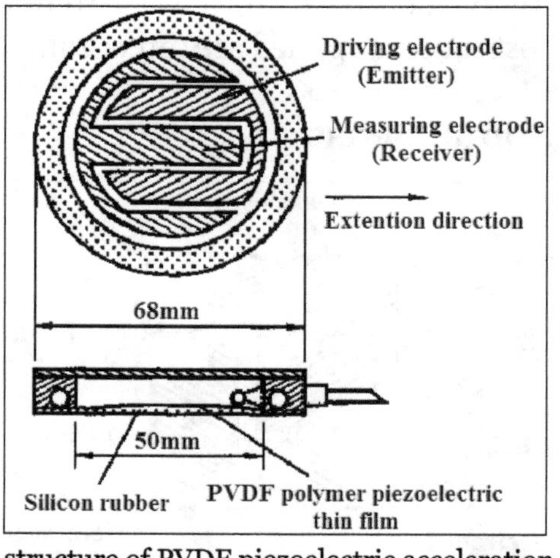

Sensing structure of PVDF piezoelectric acceleration sensor.

With the development of sensing technology, more and more heart sound sensors are appearing in our life. Although they have different sensing principle, their functions are alike.

Blood Flow Sensor

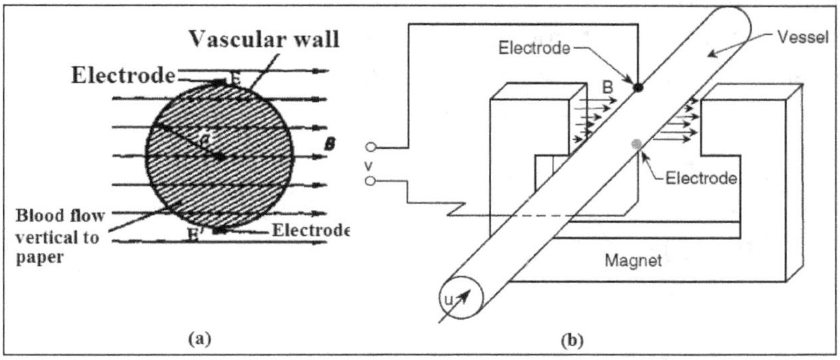

Sensing principle of electromagnetic flow sensor (a) Relation diagram between electrode and magnetic field intensity; (b) Three dimension diagram.

If oxygen and nutrients are to reach tissues, the flow of blood must be maintained in body. Cardiac output flow is often measured as an index of cardiac performance, blood flow through arterial graft is used to ensure that a graft has been successfully inserted during surgery, or the blood flow in peripheral arteries and veins may be measured to assess vascular diseases. There are usually two kinds of measuring methods: one method is direct measurement, sensor is inserted into the blood pipe to sense transient blood flow; another one is indirect measurement, sensor is placed outside vein and senses blood flow by the parameter related to the blood flow. Here, an electromagnetic flow sensor as an example is introduced to demonstrate the measurement of blood flow.

Blood flow through an exposed vessel could be measured by means of electromagnetic flow sensor. Electromagnetic flow sensor can be used in biomedicine and science research studies to measure blood flow in major blood vessels near the heart. Such sensor requires that the tested vein must be peeled off and placed into the magnetic gap of sensor. According to the output voltage of sensor, the mean velocity of vein can be calculated and known. And then in terms of the section area tested of vein, the blood flow could be gained finally.

Magnetic field intensity B is exerted along the direction vertical to vein, two electrodes are installed at both sides of vein, and then potential between two electrodes could be tested:

$$V = 2aBv_a$$

Here, B is the magnetic induction intensity at the magnetic gap; a is the radius of tested vein; υ_a is the mean velocity of vein during given test time; V is the output potential between two electrodes EE'.

Practically, this device consists of a clip-on probe that fits snugly around the blood vessel. The probe contains electrical coils to produce an electromagnetic field that is transverse to the direction of blood flow. This coil is usually excited by an AC current. A pair of very small biopotential electrodes is attached to the housing and rest against the wall of blood vessel to pick up the induced potential. The flow-induced voltage is an AC voltage at the same frequency as the excitation voltage. Utilizing AC method instead of DC excitation could help to remove any offset potential error due to the contact between the vessel wall and the biopotential electrodes.

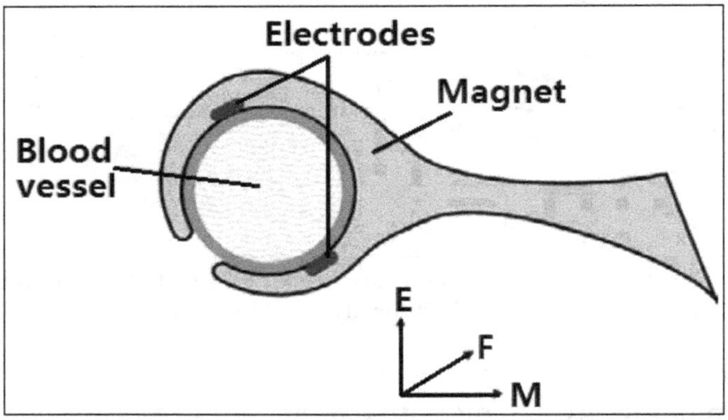

Electromagnetic blood flow sensor.

Certainly, ultrasonic wave could be also used to detect blood flow of artery. In biomedical application, there are four kinds of ultrasonic wave blood flow sensors according to specific sensing principle and methods: (1) pulse time difference; (2) voice beam deflection; (3) phase shift; (4) Doppler frequency shift.

Respiration Sensor

Respiration measurement often includes two classes: physiological parameter measurement and gas ingredient from respiration system. What sensor the former utilizes is physical sensor, and what sensor the latter employs is chemical sensor and biological sensor. Here, respiration sensor which belongs to the first class is only introduced and explained. The measurement of respiration system is important basis of clinic diagnosis, and it is necessary for patients in the fields of surgery, baby and critically ill patient's monitoring, sports medicine, and medical research. The measurement of respiration system could be classified into three classes of parameters: respiration frequency, respiration flow and lung respiration volume.

In biomedical research or clinic monitoring, respiration frequency of patient needs to be sometimes detected to record the physiological status. Figure illustrates a kind of sensor for respiration frequency based on thermistor sensing principle. Thermistor is mounted to the front-end of binder. When binder clamps the nares, airstreams from body flows through the surface of thermistor. According to the change of thermistor value, the respiration frequency would be measured.

Of course, elastic strain instrument could be also utilized to detect the respiration frequency. Its sensing principle is such: resistance wire is fixed to the surface of elastic plastic pipe. And then mercury or other electrolyte is sealed into the elastic plastic pipe. After elastic plastic pipe is adhered to the front of breast, respiration would lead to the length change of elastic plastic pipe. Such length change causes the change of resistance wire which could show the change of respiration frequency. When resistance wire is introduced into a detecting circuit, the respiration frequency will be sensed and measured.

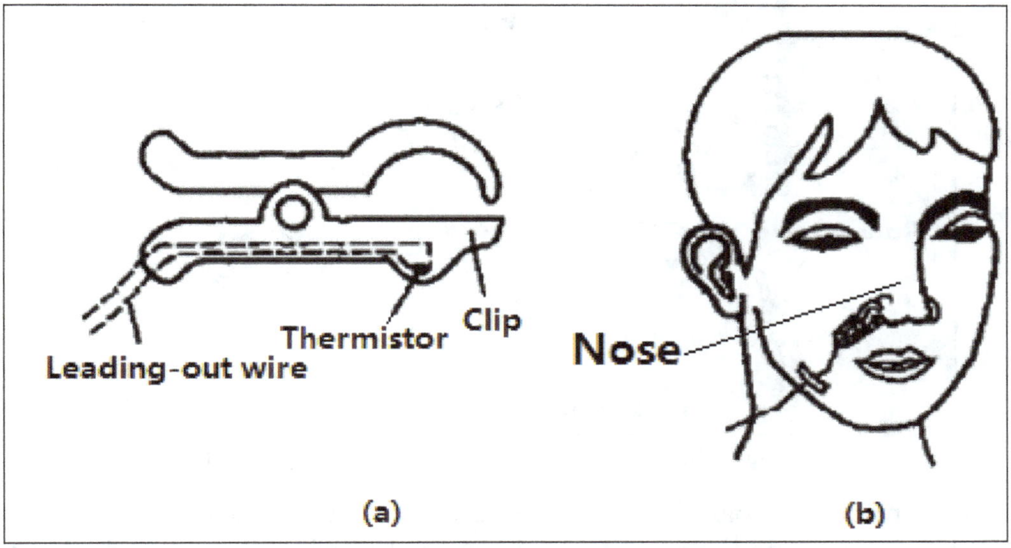

Thermistor sensor for respiration frequency
(a) structure diagram, (b) measurement diagram.

Blood Pressure Sensor

If blood circulation is to be maintained in the body, tissues are to be perfused with oxygen. Then correct pressure measurement has to be applied in the vascular system. The usual blood pressure methods have: liquid coupling direct measurement, pipe-end sensing measurement, indirect blood pressure sensing measurement. Liquid coupling direct measurement means that the pipe filled with liquid is inserted into the measured part and that the pressure is measured by liquid coupling of pipe end position in the body, which is the simplest method. Pipe-end sensing measurement employs pipe-end sensor to measure blood pressure. Pipe-end sensor which can convert the pressure signal into electronic signal is placed on the measured part. And then the electronic signal measured is transmitted to the external wire. Such method could avoid the distortion of signal of blood pressure. Pipe-end sensing measurement has a lot of advantages, but such method needs to activate the skin and relative sensors have to been placed into the body. Hence indirect blood pressure measurement is noted by people and continuously explored. Blood-pressure meter is a classic example of indirect blood pressure measurement, which is shown in figure.

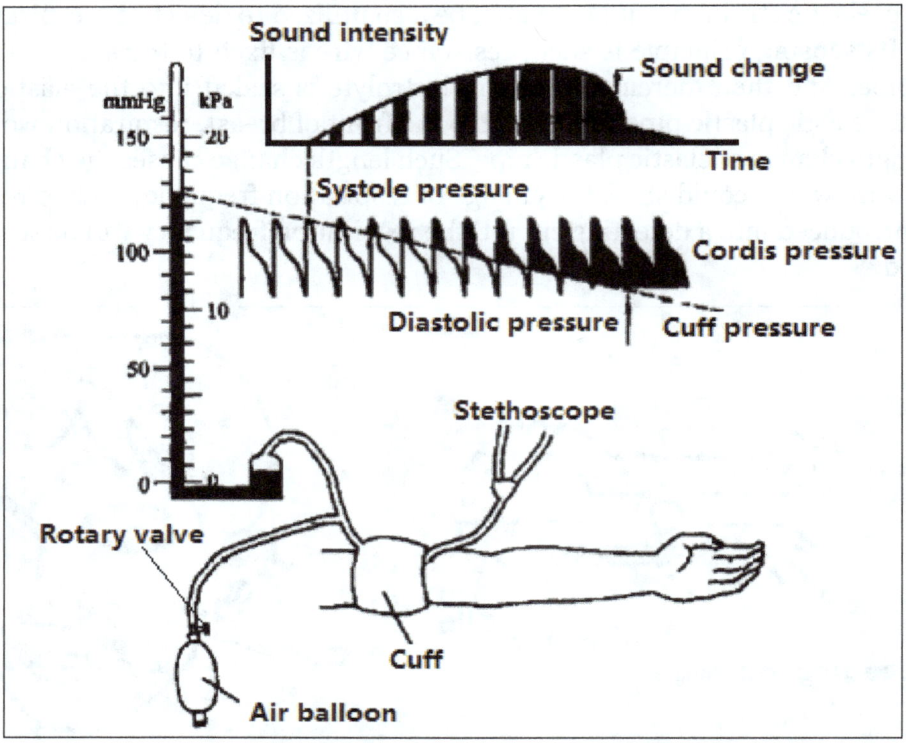

Sensing principle of indirect blood pressure measurement.

The sensing principle is based on Coriolis sound. Gas is filled into cuff to hold back the arterial blood flow. And then gas in cuff is sent out slowly to monitor whether there appears arterial blood flow at the downstream of arterial blocking point. Here, the employed sensor is common mercury pressure meter. And such method is up to the actual experience of staff. When coriolis sound is heard, namely when blood flows through artery blood pipe, blood pressure in cuff is shrinking pressure in artery pipe. When blood flows recover normal level, blood pressure in cuff is diastolic pressure of artery. Systole pressure and diastolic pressure are recorded as blood pressure. Such method is not harmful to the skin or organ in the body.

Electrochemical Electrode

Biopotential measurements are made using different kinds of specialized electrochemical electrodes. The function of electrodes is to couple the ionic potentials generated inside the body to an electronic instrument. Biopotential electrochemical electrode is classified either as noninvasive (e.g. skin surface) or invasive (e.g. microelectrode, wire electrode) electrodes. When a metal is placed in an electrolyte solution, a charge distribution is created next to the metal/electrolyte solution interface as illustrated in figure. The localized charge distribution causes an electronic potential by electrochemical electrode, called half-cell potential, to be developed across the interface between metal electrode and electrolyte solution.

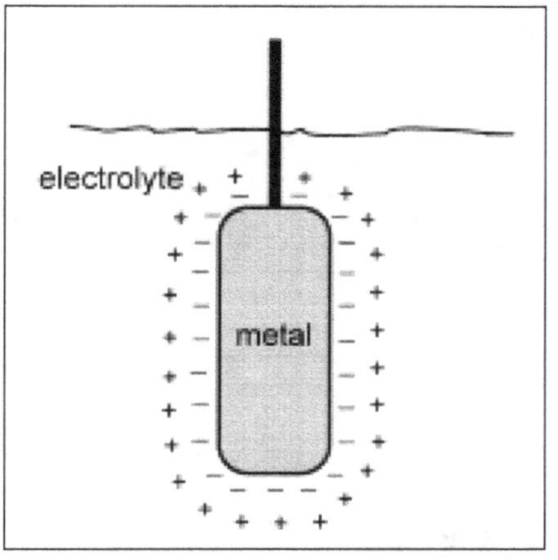

The charge distribution at a electrolyte/metal interface.

The half-cell potentials of several important metals are listed in table. Here, a point needs to be pointed out that hydrogen electrode is considered to be a standard electrode against which the half-cell potentials of other metal electrodes are measured.

Silver and zinc electrodes are immersed in an electrolyte solution. And then we may calculate the potential drop between two electrodes. From table, the half-cell potentials for silver and zinc electrodes are 0.799V and -0.763V respectively. Hence, the half-cell potentials between two electrodes are equal to the following value:

$$0.799-(-0.763)=1.562V$$

Typically, utilizing the electrochemical electrodes composed of the same metals could measure the half-cell potentials. Hence, the two half-cell potentials for these electrodes would be equal in magnitude. Some common electrodes are introduced here, which is utilized as a sensor.

ECG Electrodes

A typical flexible biopotential electrode for ECG (electrocardiogram, ECG) recording is composed of certain polymers or elastomers which are made electrically conductive by the addition of a fine carbon or metal powder. These electrodes as illustrated in figure (a) are available with prepasted AgCl gel for quick easy application to the skin using a double-sided peel-off adhesive tape. The most common type of biopotential electrode is the silver/silver chloride electrode (Ag/AgCl), which is formed by electrochemically depositing a very thin layer silver chloride onto the surface of silver electrode as illustrated in figure (b). These electrodes are recessed and imbedded in foam that has been soaked with an electrolyte paste to provide good electrical contact with the skin. The electrolyte saturated foam is also known to reduce motion artifacts which are produced during stress testing when the layer of the skin moves relative to the surface of the Ag/AgCl electrode. This motion leads to the large interference in the recorded biopotential and, in the extreme cases, could severely degrade the measurement.

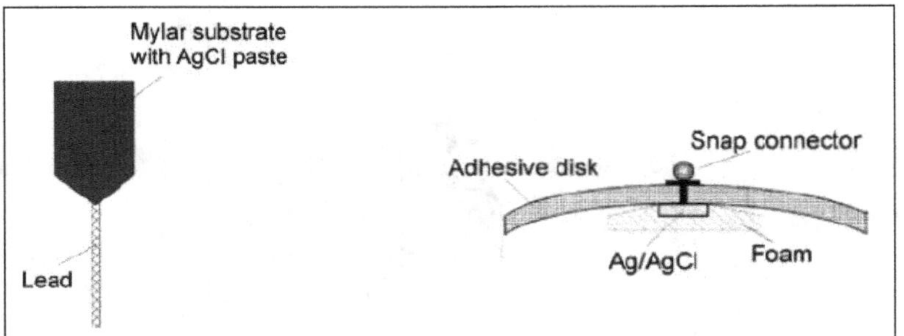

(a) Flexible Mylar electrode (b) Disposable snap-type Ag/AgCl electrode

Biopotential skin surface ECG electrode.

Table: Half-cell Potentials of Important Metals.

Primary Metal and Chemical Reaction	Half-cell Potential
$Al \rightarrow Al^{3+} + 3e^-$	-1.706
$Cr \rightarrow Cr^{3+} + 3e^-$	-0.744
$Cd \rightarrow Cd^{2+} + 2e^-$	-0.401
$Zn \rightarrow Zn^{2+} + 2e^-$	-0.763
$Fe \rightarrow Fe^{2+} + 2e^-$	-0.409
$Ni \rightarrow Ni^{2+} + 2e^-$	-0.230
$Pb \rightarrow Pb^{2+} + 2e^-$	-0.126
$H_2 \rightarrow 2H^+ + 2e^-$	0.000(stand by definition)
$Ag \rightarrow Ag^+ + e^-$	+0.799
$Au \rightarrow Au^{3+} + 3e^-$	+1.420
$Cu \rightarrow Cu^{2+} + 2e^-$	+0.340
$Ag + Cl \rightarrow AgCl + 2e^-$	+0.223

EMG Electrodes

Electrochemical electrodes are also used to record electromyography (EMG) signals from different muscles in the body. The body and size of the recorded EMG signals depends on the electrical property of these electrodes and the recording location. For invasive recordings, proper skin preparation, which normally involves cleaning the skin with alcohol or the application of a small amount of an electrolyte paste, helps to minimize the impedance of the skin-electrode interface and to improve the quality of recording signal considerably. The most common electrodes used for the surface EMG recording and nerve conduction studies are circular discs, about 1 cm in diameter, that are made of silver or platinum. For direct recording of electrical signals from nerves and muscle fibers, a variety of percutaneous needle electrodes are available as illustrated in figure. The most common type of needle electrode is the concentric bipolar electrode as illustrated in figure (a). This electrode is made from the thin metallic wires encased inside a larger canola or hypodermic needle. The two wires serve as the recording and reference electrodes.

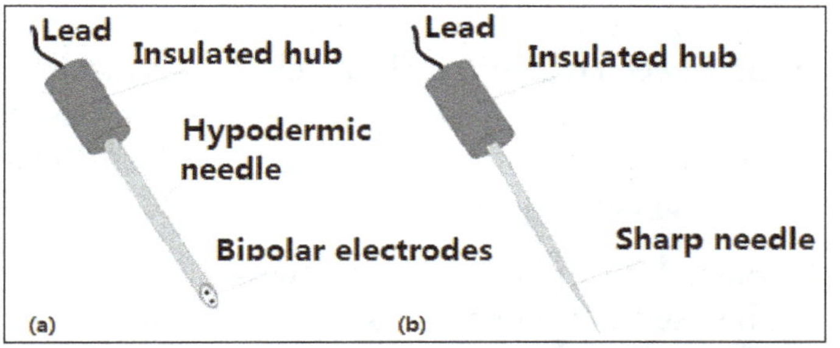

Intramuscular biopotential electrode: (a) bipolar electrode, (b) unipolar configuration.

Another type of percutaneous EMG electrode is the unipolar needle electrode as illustrated in figure (b). This electrode is made of a thin wire that is most insulated by a thin layer near the distal tip. Unlike bipolar electrode, this type of electrode requires a second unipolar reference electrode to form a closed electrical circuit. The second recording electrode is normally placed either adjacent to the recording electrode or attached to the surface of our skin.

EEG Electrodes

The most commonly used electrode for recording electroencephalographic (EEG) signals from the brain are cup electrodes and subdermal needle electrodes. Cup electrodes are made of platinum or tin and are approximately 5-10mm in diameter. The cup electrodes are filled with an electrolyte gel and can be attached to the scalp with an adhesive tape.

Recording the biopotentials from the scalp is very difficult because hair and oily skin hold back the good electrical contact. Hence, clinicians sometimes prefer to use subdermal needle electrodes (EEG electrodes) instead of the metal surface electrodes for EEG recording. These electrodes are both fine platinum or stainless-steel needle electrodes about 10mm long by 0.5mm wide, which are inserted under the skin to provide a better electrical contact.

Microelectrodes

Microelectrodes are biopotential electrodes with ultra-fine tapered tip that can be inserted into biological cells. These electrodes play a very important role in recording action potentials from single cells and are used in neurophysiologic studies to comprehend the course of biological information conversion and transmission in our body. The tip of these electrodes must be very small with respect to the dimensions of the biological cell to avoid cell damage and at the same time sufficiently strong to penetrate the cell wall. The electrode which is applied to microbe studies is called microelectrodes. Generally, there are three types of microelectrodes: (1) glass microelectrodes, (2) metal electrodes, and (3) solid-state microprobes.

For glass microelectrodes, when the tip of such electrodes is inserted into an electrolyte solution, such as the intracellular cytoplasm of a biological cell, ionic current can flow through the fluid junction at the tip of the microelectrode. Such mode could establish a closed electrical circuit between two Ag/AgCl wire electrodes inside the microelectrode and biological cell. For metal electrode, when the tip of such microelectrodes is usually sharpened down to a diameter of a few micrometers by an electrochemical etching process. The wires are then insulated up to its tip.

Solid-state microfabrication techniques commonly used in the production of the integrated circuits can be used to produce microprobes for multichannel recordings of biopotentials or for electrical stimulation of neurons in our brain or spinal cord. Most of solid-state microelectrodes are microsensor actually. Such probe consists of a precisely micromachined silicon substrate with four exposed recording sites. One of main advantages of microfabrication techniques is the ability to mass produce very small and highly sophisticated microsensors with highly reproducible electrical and physical properties.

Enzyme Sensor and Microbial Sensor

Enzyme constitutes a group of more than 2000 proteins having so-called biocatalytic properties. These properties give the enzymes the unique and powerful ability to accelerate chemical reactions inside biological cells. Most enzymes react only with specific substrates even though they can be contained in a complicated mixture with other substances. It is important that soluble enzymes are very sensitive both to temperature and pH variations, and they can be inactivated by many chemical inhibitors. For practical biosensor applications, these enzymes are normally immobilized by insolubilizing the free enzymes via entrapment into an inert and stable matrix such as starch gel, silicon rubber, or polyacrylamide. This process is very important to ensure that the enzymes retains its catalytic properties and can be reusable.

The action of specific enzymes may be utilized to form a range of different biosensors. A typical example of enzyme-based sensor is a glucose sensor that uses the enzyme glucose oxidase. Glucose plays an important role in metabolic process. Currently, available glucose sensors are based on an immobilized enzyme, such as glucose oxidase, which acts as a catalyst. Glucose is detected by electromechanically measuring either the amount of gluconic acid or hydrogen peroxide (H_2O_2) produced or by measuring the percent of oxygen consumed according to the following chemical reaction:

$$\text{Glucose} + O_2 + H_2O \xleftrightarrow{\text{glucose oxidase}} \text{gluconic acide } H_2O_2$$

A glucose sensor is similar to a pO2 sensor and is shown in figure. Glucose and oxygen enter through outer membrane to interact with glucose oxidase enzyme. The remaining oxygen penetrates through the second oxygen-permeable membrane and is measured by the oxygen electrode.

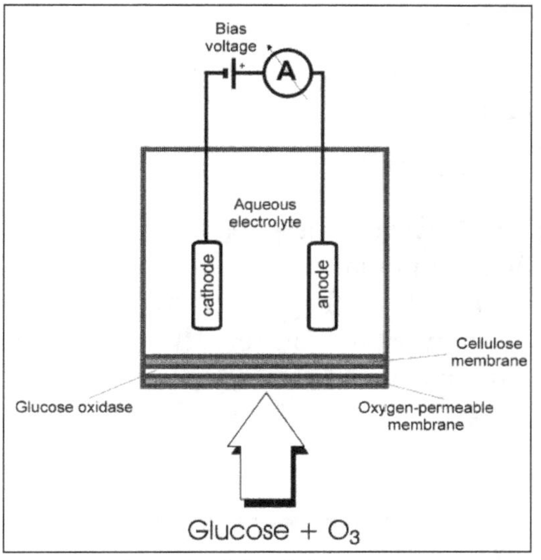

Sensing principle of glucose sensor.

Biocatalytic enzyme-based sensors generally consist of an electrochemical gas-sensitive converter or an ion-selective electrode with an enzyme immobilized in or on a membrane that serve as the biological mediator. The analyte diffuses from the bulk sample solution into the biocatalytic layer where an enzymatic reaction takes place. The electroactive product that is formed (or consumed) is usually detected by an ion-selective electrode. A membrane separates the basic sensor from the enzyme if a new gas is produced (such as CO_2 or NH_3) or consumed (such as O_2). Although the concentration of the bulk substrate drops continuously, the rate of consumption is usually negligible. The decrease is detected only when the test volume is very small or when the area of enzyme membrane is large enough. Thus this electrochemical analysis is nondestructive, and the sample is reused. Measurements are usually performed at a constant pH and temperature either in a stirred medium solution or in a flow through solution. In order to control biochemical process including some enzyme sensors, a number of microbial sensors have been continuously developed and applied to various environment, agriculture, food and pharmaceutical.

Microbial sensors typically involve the assimilation of organic compounds by microorganisms, followed by a change in respiration activity(metabolism) or the production of specific electrochemically active metabolites, such as CO_2, H_2 or NH_3, that are secreted by the microorganism.

A microbial sensor is composed of immobilized microorganisms that serve as specific recognition elements and an electrochemical or optical sensing device that is used to convert the biochemical signal into electronic signal that can be processed. The operation of a microbial sensor can be described by the following five-step process:

- The substrate is transported to the surface of the sensor.
- The substrate diffuses the membrane to the immobilized microorganisms.

- A reaction occurs at the organism.
- The products formed in the reaction are transported through the membrane into the surface of detector.
- The product is measured by the detector.

Here, an example of a microbial sensor is given to demonstrate the detecting course of microbial sensor including ammonia (NH_3) and nitrogen dioxide (NO_2) sensors that utilize the nitrifying bacteria as the biological sensing component. A NH_3 biosensor can be constructed on the base of nitrifying bacteria that uses ammonia (NH_3) as a source of energy and oxidizes ammonia as follows:

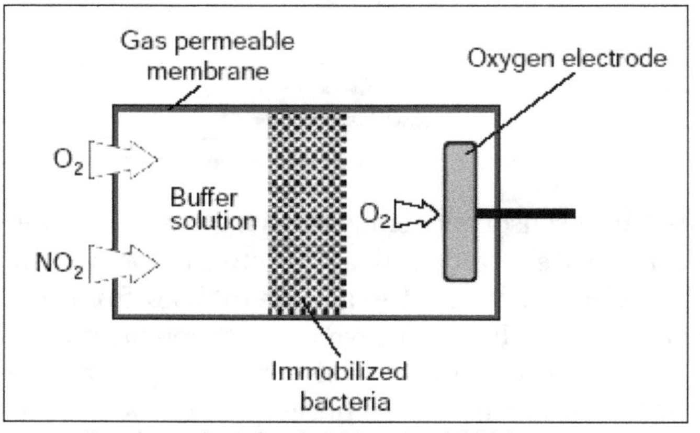

Sensing principle of a NO_2 microbial biosensor.

$$NH_3 + 1.5O_2 \xrightarrow{\text{Nitrosomonas}} NO_2 + H_2O + H^+$$

This oxidation process proceeds at high rate, and the amount of oxygen consumed by the immobilized bacteria can be measured directly by a polarographic oxygen electrode placed behind the bacteria.

Nitric oxide (NO) and NO_2 are two principal pollution gases of nitrogen in the atmosphere. The principle of a NO_2 biosensor is shown in figure. When a sample of NO_2 gas diffuses through the gas-permeable membrane, it is oxidized by the nitrosomonas bacteria as follows:

$$2NO_2 + O_2 \xrightarrow{\text{Nitrosomonas}} NO_3$$

Similar to an ammonia biosensor, the consumption of oxygen O_2 around the membrane is determined by an electrochemical oxygen electrode.

Biomeasurement System and Instrumentation

Biomeasurement System Constitution

The biomeasurement system is shown in figure to measure some biological signals such as quantity, property, or condition which are bioelectrical signal generated by muscles or the brain, or a chemical or mechanical signal that is converted into an electrical signal. Biomeasurement system is composed of sensor, analog processing circuit, A/D conversion, digital signal processing, output display, and data storage. A/D conversion is used in bioinstrumentation to acquire the enough system gain.

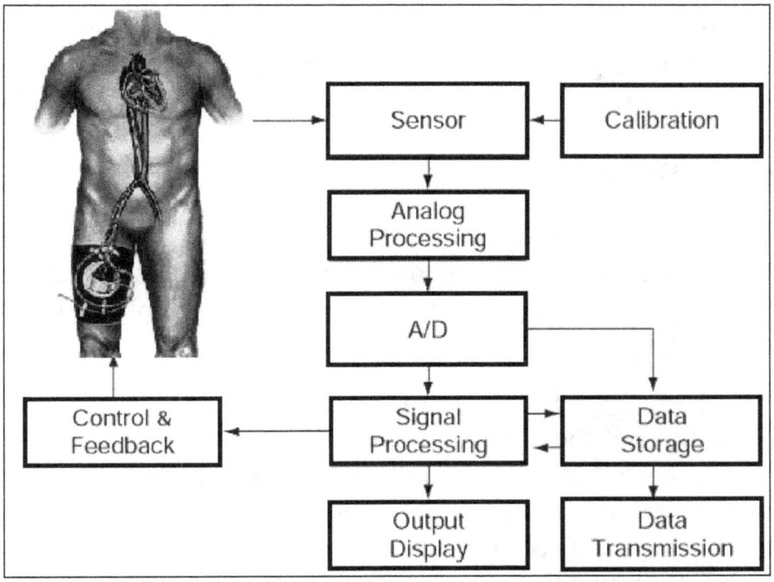

Basic biomeasurement system using sensors to measure a biological signal with data acquisition, storage and display capabilities, data transmission, along with control and feedback.

With the invention of telephone and with appearance of internet, signal can be required with a device in one location, perhaps in patient's home, and transmitted into another device for transmission or storage. For example, if a biological signal from bioinstrumentation system in rural area could be transmitted into a diagnosis center in hospital, doctor would quickly judge some diseases to make patients gain an accurate treatment or diagnosis in time.

Two other components play important roles in bioinstrumentation system. The first is the calibration signal. A signal with known frequency and amplitude is applied to the bioinstrumentation system at sensor's input. The calibration device permits the system components to be adjusted so that it's known that the output and input have a certain linear relationship. Without such information, it's impossible to convert the output of an instrument system into a meaningful representation of the measurand.

Another important component, a control or feedback element, is not a part of all instrument systems. These parts include pacemakers and ventilators that could stimulate the heart and lungs. Some feedback devices collect physiological data and stimulate a response— a beat or breath—when needed or are part of biofeedback systems in which patients are made aware of a physiological instrument, such as blood pressure, and uses conscious control to change the physiological response.

Biomeasurement Circuit

The basic elements which constitute basic bioinstrumentation system. Circuits play a very important role in bioinstrumentation system. If a bioinstrumentation needs to be developed or improved to be fit for new condition, function circuits in different blocks to be respectively designed to form bioinstrumentation system with relative indexes. Among all kinds of circuits, amplifiers and A/D converters are very important component for detecting the biological signal. Hence, amplifying circuits will be only introduced here in detail, and other function circuits could be read or utilized in relative books.

Bioelectric Amplifier

In order to record the bioelectric potential from the body, biological amplification is always required. The simplest form is shown in figure which uses a single-input amplifier. Here amplifier only amplifies one input signal which is applied in the input and the reference "earth" or "ground".

In this amplifier, the resistor R_1 is required to allow the "bias current" to flow into the non-inverting (+) input of the operational amplifier and the resistor R_2 is required to balance the resistor R1 so that the bias currents do not produce a voltage difference between the two inputs of the amplifier. If there is no capacitor in the positive input, amplifier in figure will become a voltage follower, that's to say, the bioelectric input signal could be transmitted completely to the output. Namely, the following equation is given:

$$u_{in} + = u_{out}$$

Unfortunately, in circuit shown in figure, the resistor R_1 defines the maximum input impedance of the amplifier. The input impedance is an important consideration in bioelectric amplifiers because it can cause attenuation of a signal which is derived from electrodes with high impedances. For example, if the two electrode impedances were 10k Ω and the input impedance of the amplifier was 1M Ω, then 1% of the signal would be lost by attenuation of two electrodes. The impedance presented by electrodes is termed the source impedance which has to be very much less than the input impedance of amplifier. Source impedance is very important when we consider differential amplifier shortly.

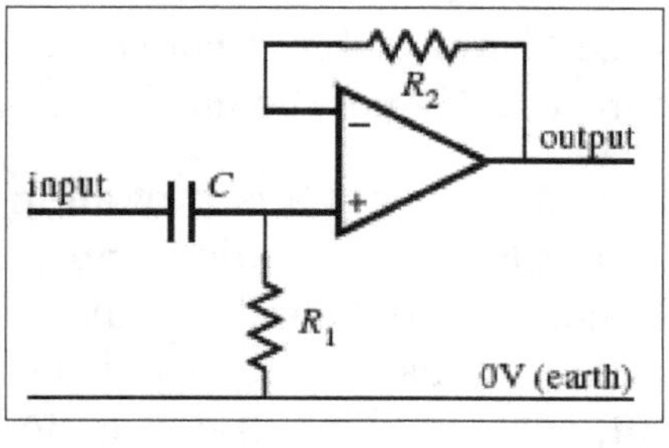

A simplest single-input amplifier.

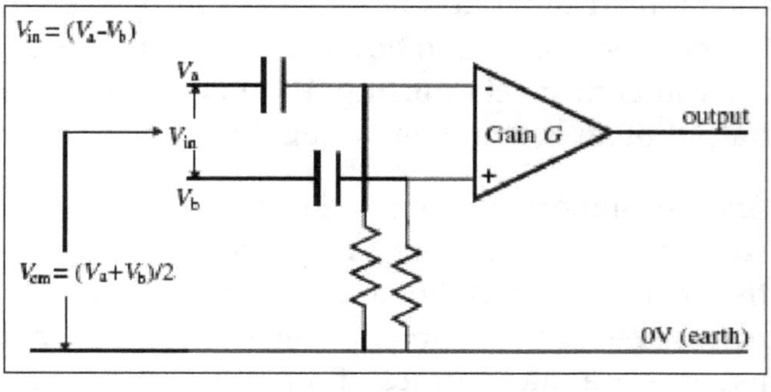

A differential amplifier.

There is also capacitor introduced in figure in series with the input signal. Capacitor C blocks any DC (direct current) signal by acting as a high-pass filter with the resistor R_1. This function is usually referred as AC (alternative current) coupling. AC coupling will also cause some attenuation of the signal which may be important. We could determine the attenuation of signal by the following equations:

$$u_{out} = u_{-} = u+, \frac{u_{in}}{R1 + \frac{1}{j\omega C}} = \frac{u+}{R1}, \text{ then } \frac{u_{out}}{u_{in}} = \frac{R1}{R1 + \frac{1}{j\omega C}} = \frac{R1(R1 + j/\omega C)}{R_1^2 + 1/\omega^2 C^2}$$

$$\left|\frac{U_{out}}{U_{in}}\right| = \frac{1}{\sqrt{1 + 1/(\omega^2 R_1^2 C^2)}} = \frac{1}{\sqrt{1 + \omega_o^2/\omega^2}}, \text{where } \omega_o = 1/R_1 C.$$

If capacitor C is 1μF and resistor R_1 is 1MΩ, then the attenuation of a 1Hz signal will be 1.25%. This is perhaps a significant attenuation for ECG which has considerable energy at 1Hz.

Unfortunately, even with capacitor C added, this type of amplifier is not suitable for recording small bioelectric signal because of interference from external electric fields. An electric electrode has to be connected to the amplifier via a wire and this wire is exposed to interfering signals. However, the interference will only appear on the input wire to the amplifier and not on the "ground" wire which is held at zero potential. An elegant solution to this problem is to utilize differential amplifier. The input of the type of amplifier has three connections marked '+', '-' and 'ground'.

Differential Amplifier

In figure, the signal which we wish to record is connected between the '+' and '-' points. Now both inputs are exposed to any external interfering electric fields so that the difference between the '+' and '-' input will be zero. This will not be quite true because the electric fields experienced by the two input wires may not be exactly the same, but if the wires are run close together then the difference will be small. Differential amplifier is not perfect in that even with the same signal applied to both inputs, with respect to ground; a small output signal can appear. This imperfection is specified by the common mode rejection ratio or CMMR. An ideal differential amplifier has zero output when identical signals are applied to the two '+' and '-' inputs. CMMR could be defined as the following equation:

$$\text{CMMR} = 20\log\left[\frac{\text{signal gain}}{\text{common} - \text{mode gain}}\right]$$

Where, signal and common-mode gains are given by:

$$\text{signal gain} = \frac{U_{out}}{U_{in}} = \frac{U_{out}}{U_{+} - U_{-}}$$

$$\text{common} - \text{mode gain} = \frac{U_{out}}{U_{cm}} = \frac{U_{out}}{(U_a + U_b)/2}$$

In practice, common-mode voltage U_{cm} can be as large as 100mV or even more. In order to reject this signal and record a signal V_{in} as small as 100μV, a high CMMR is required. If we wish the interfering signal to be reduced to only 1% of output voltage then:

$$\text{required} - \text{signal gain} = \frac{U_{out}}{U_{in}} = \frac{U_{out}}{U_+ - U_-} = \frac{U_{out}}{100\mu V}$$

$$\text{required} - \text{CM gain} = \frac{U_{out}}{U_{cm}} = \frac{U_{out}}{1\% U_{in}} = \frac{U_{out}/100}{100mV}$$

Hence, the required CMMR could be given as:

$$\text{CMMR} = 20\log\left[\frac{U_{out}/0.1mV}{U_{out}/100^2 mV}\right] = 100dB$$

In fact, it's not always easy to achieve a CMMR of 100dB. As we have known, electrode source impedances have a very significant effect on CMMR and hence electrode impedance affects noise rejection.

Of course, in detecting biosignals, the AC coupling shown in figure and figure degrades the performance of the amplifiers. If the input impedance and bias current of amplifiers is sufficiently high, then they can be connected directly to the input electrodes, without producing electrode polarization. Furthermore, DC offset will occur from the electrode contact potentials, but if the amplifier gain is low (<10) DC offset will be not a significant problem. The offset can be removed by AC coupling at later stage.

However, there are some safety arguments against the use of DC coupling. If a fault arises in the operational amplifier, then it's possible for the power supply to be directly connected to the patient and so give rise to a hazard. DC currents will cause electrolysis and result in tissue necrosis. AC coupling could avoid this problem and is often used. Nonetheless DC coupling is also often used in biomedical field.

Bioinstrumentation Design

The purpose of using bioinstrumentation is to monitor the output of a sensor or sensors and to extract some useful information from signals that are produced by sensors.

Acquiring discrete-time signal and storing this signal in computer memory from a continuous-time signal is accomplished with analog-to-digital (A/D) converter. After analog signals have been processed which are based on analog filters such as low-pass or high-pass filters, A/D converter uniformly samples the continuous-time waveform and transforms it into a sequence of numbers, one every tk seconds. The A/D converter also transforms the continuous-time waveform into a digital signal, which is converted into computer words and stored in computer memory. To adequately capture the continuous-time signal, the sample frequency has to be carefully selected to ensure any signal information is not lost. The minimum sampling frequency is twice the highest frequency content of the signal based on the sampling theorem from communication theory. In reality, we often adopt the sampling frequency from five to ten times the highest frequency content of the signal so as to achieve better accuracy by reducing aliasing error.

Biological Signal Categories in Human Body

The electrical, chemical and mechanical activity that occurs during this biological event often produces signals that could be detected and analyzed. Biological signals are the record of a biological event such as a beating heart or a contracting muscle. Hence, biological signals contain useful information which could reflect human's activities and physiology, that's to say, biological signal could be used for biomedical diagnosis. Biological signals are classified into bioelectric signals, biomagnetic signals, biochemical signals, biomechanical signals, bioacoustic signals and biooptical signals.

Nerve and muscle cells generate bioelectric signals that are the result of electrochemical changes within and between cells. When plenty of cells are stimulated, an electric field is then generated that propagates through biological tissues. These changes in extracellular potential may be measured on the surface of tissue or organism by using surface electrodes. The electrocardiogram (ECG) is an example of this phenomenon. Different organs in body, including the heart, brain, lungs, and liver, also generate weak magnetic fields that could be detected with magnetic sensors. The strength of magnetic field is much weaker than the corresponding physiological bioelectric signal. Magnetic sensors could be used to detect biomagnetic signals. Magnetocardiography (MCG) is a specific example of such phenomenon.

Biochemical signals contain information about changes in the concentration of various chemical agents in the body. The concentration of various ions such as calcium and potassium in cell can be measured and recorded. Oxygen sensor is used to detect oxygen concentration in body. Mechanical functions of biological systems, including motion, displacement, tension, force, pressure and flow, also produce measurable biological signals. Blood pressure sensor is a measurement of the force that blood exerts against the walls of blood vessels. Change in blood pressure can be recorded as a waveform by blood pressure sensor. Bioacoustics' signals are a special subset of biochemical signals which involve vibrations. Many biological events could produce acoustic noise. For example, the flow of blood through the valves in the heart can be used to determine whether motion is operating properly. Besides these, the respiratory system, joints and muscles could also produce bioacoustic signals that propagate through the biological medium and can be often measured at the skin surface by acoustic sensors. Biooptical signals are generated by the optical or light induced attributes of biological systems. Biooptical signals can occur or be introduced to measure a biological parameter with an external light medium such as the measurement of health of a fetus by red and infrared light.

Noise

Measurement signals are always corrupted by noise in the bioinstrumentation system. Interference noise occurs when unwanted signals are introduced into systems by external sources such as telephone magnetic wave, power line and transmitted radio. Interference noise needs to be effectively reduced by careful attention to the circuit wiring configuration to minimize coupling effect.

Interference noise is introduced by power lines, fluorescent lights, AM/FM radio broadcasts, computer clock oscillator, laboratory equipment and cellphone. Electromagnetic energy radiating from noise source is injected into the amplifier circuit or into the patient by capacitive or inductive coupling. Even action potentials from nerve conduction in the patient generate noise at the sensor/amplifier surface. Filters are also used to reduce the noise and to maximize the signal-to-noise(S/N) rate at the input of the A/D converter.

Low frequency noise could be eliminated by high-pass filter with the cutoff frequency set above the noise frequency. High frequency noise could be reduced by low-pass filter with the cutoff frequency set below the noise frequency and above the frequency of biological signal which is being monitored. Power line noise is a very difficult problem in biological monitoring because the 50-or-60-Hz frequency is usually at the range of biological signal which could be monitored. Band-stop filters are commonly used to reduce the power line noise. The notch frequency in the band-stop filters is set to the power line frequency of 50 or 60Hz with the cutoff frequency located a few Hertz to either side.

The second type of noise is called inherent noise. Inherent noise arises from random processes that are fundamental to the operation of circuit's elements and thus is reduced by a good circuit design practice. While inherent noise is reduced, it can be never eliminated. Low-pass filters are used to reduce high-frequency components. However, noise signals within the frequency range of biological signals being amplified cannot be eliminated by this filtering approach.

Computer

Computer is a main device which is used to display the biological signals being monitored. However some low or high level languages such as machine language, FORTRAN, visual C++, MATLAB or LabView, have to be used to realize the operation on the acquisition data from biological body. When computers are used to acquire physiological data, programming instruction tell computer when acquisition data should begin, how often samples should be taken from how many sensors, how long acquisition data should continue, and where the digitized data should be stored. The rate at which a system acquires sample depends on the speed of computer clock's frequency and the number of computer instruction that could be completed in order to realize a sample. Of course, some computers are utilized to control the gain on the input amplifiers so that biological signals could be adjusted during data acquisition. In other systems, the gain of data acquisition has to be adjusted.

References

- Ota K (2010). "Advances in artificial lungs". Journal of Artificial Organs. 13 (1): 13–16. doi:10.1007/s10047-010-0492-1. PMID 20177723
- Biomedical-time-series-processing-and-analysis-methods-the-case-of-empirical-mode-decomposition, advanced-biomedical-engineering: intechopen.com, Retrieved 18 February, 2020
- Semino CE; Kasahara J; Hayashi Y; Zhang S (2004). "Entrapment of migrating hippocampal neural cells in three-dimensional peptide nanofiber scaffold". Tissue Eng. 10 (3–4): 643–55. doi:10.1089/107632704323061997. PMID 15165480
- What-is-biomechanics, faq: canyonsportstherapy.com, Retrieved 16 April, 2020
- Kurzmann, Ernst; Fladerer, Johannes-Paul (2017). ManagemANT Was Fach- und Führungskräfte von Ameisen lernen können (1. Auflage ed.). Frankfurter Allgemeine Buch. ISBN 9783956012082
- Biomedical-sensor-device-and-measurement-systems, advances-in-bioengineering, book: intechopen.com, intechopen.com, Retrieved 27 June, 2020
- Hetling, JR (2008). ""Comment on "What is Neural Engineering". Journal of Neural Engineering. 5 (3): 360–1. doi:10.1088/1741-2560/5/3/n01. PMID 18756032

3

Medical Biology

The branch of biology that deals with the practical applications of medicine, health care and laboratory diagnostics is called medical biology. It includes other disciplines such as molecular biology, nanobiotechnology, cytogenetics, microbiology, toxicology, etc. This chapter discusses the subject of medical biology in detail.

Medical biology is a field of biology that has practical applications in medicine, health care and laboratory diagnostics. Medical biology is based on molecular biology that combines all issues of developing molecular medicine into large-scale structural and functional relationships of the human genome, transcriptome, proteome and metabolome with the particular point of view of devising new technologies for prediction, diagnosis and therapy.

PATHOLOGY

Pathology is the study of disease by scientific methods. 'Patho' means disease and 'logy' means study, therefore pathology is a scientific study of disease. Diseases may, in turn, be defined as an abnormal variation in structure or function of any part of the body. Pathology gives explanations of a disease by studying the following four aspects of the disease:

- Etiology,
- Pathogenesis,
- Morphologic changes and
- Functional derangements and clinical significance.

Etiology

Etiology of a disease means the cause of the disease. If the cause of a disease is known it is called primary etiology. If the cause of the disease is unknown it is called idiopathic. Knowledge or discovery of the primary cause remains the backbone on which a diagnosis can be made, a disease understood, & a treatment developed. There are two major classes of etiologic factors: genetic and acquired (infectious, nutritional, chemical, physical, etc).

Pathogenesis

Pathogenesis means the mechanism through which the cause operates to produce the pathological and clinical manifestations. The pathogenetic mechanisms could take place in the latent or incubation period. Pathogenesis leads to morphologic changes.

Morphologic Changes

The morphologic changes refer to the structural alterations in cells or tissues that occur following the pathogenetic mechanisms. The structural changes in the organ can be seen with the naked eye or they may only be seen under the microscope. Those changes that can be seen with the naked eye are called gross morphologic changes & those that are seen under the microscope are called microscopic changes. Both the gross & the microscopic morphologic changes may only be seen in that disease, i.e. they may be specific to that disease. Therefore, such morphologic changes can be used by the pathologist to identify (i.e. to diagnose) the disease. In addition, the morphologic changes will lead to functional alteration & to the clinical signs & symptoms of the disease.

Functional Derangements and Clinical Significance

The morphologic changes in the organ influence the normal function of the organ. By doing so, they determine the clinical features (symptoms and signs), course, and prognosis of the disease.

Etiology →Pathogenesis →Morphologic changes →Clinical features & Prognosis of all diseases.

Understanding of the above core aspects of disease (i.e. understanding pathology) will help one to understand how the clinical features of different diseases occur & how their treatments work. This understanding will, in turn, enable health care workers to handle & help their patients in a better & scientific way. It is for these reasons that the health science student should study pathology. In addition, the pathologist can use the morphologic changes seen in diseases to diagnose different diseases. There are different diagnostic modalities used in pathology. Most of these diagnostic techniques are based on morphologic changes.

Diagnostic Techniques used in Pathology

The pathologist uses the following techniques to the diagnose diseases:

- Histopathology.
- Cytopathology.
- Hematopathology.
- Immunohistochemistry.

- Microbiological examination.
- Biochemical examination.
- Cytogenetics.
- Molecular techniques.
- Autopsy.

Histopathological Techniques

Histopathological examination studies tissues under the microscope. During this study, the pathologist looks for abnormal structures in the tissue.

Tissues for histopathological examination are obtained by biopsy. Biopsy is a tissue sample from a living person to identify the disease. Biopsy can be either incisional or excisional.

Once the tissue is removed from the patient, it has to be immediately fixed by putting it into adequate amount of 10% Formaldehyde (10% formalin) before sending it to the pathologist. The purpose of fixation is:

- To prevent autolysis and bacterial decomposition and putrefaction.
- To coagulate the tissue to prevent loss of easily diffusible substances.
- To fortify the tissue against the deleterious effects of the various stages in the preparation of sections and tissue processing.
- To leave the tissues in a condition which facilitates differential staining with dyes and other reagents.

Once the tissue arrives at the pathology department, the pathologist will exam it macroscopically (i.e. naked-eye examination of tissues).

Then the tissue is processed to make it ready for microscopic examination. The whole purpose of the tissue processing is to prepare a very thin tissue (i.e. five to seven μm or one cell thick tissue) which can be clearly seen under the microscope. The tissue is processed by putting it into different chemicals. It is then impregnated (embedded) in paraffin, sectioned (cut) into thin slices, & is finally stained. The stains can be Hematoxylin/Eosin stain or special stains such as PAS, Immunohistochemistry, etc. The Hematoxylin/Eosin stain is usually abbreviated as H&E stain. The H&E stain is routinely used. It gives the nucleus a blue color & the cytoplasm & the extracellular matrix a pinkish color. Then the pathologist will look for abnormal structures in the tissue. And based on this abnormal morphology he/she will make the diagnosis. Histopathology is usually the gold standard for pathologic diagnosis.

Cytopathologic Techniques

Cytopathology is the study of cells from various body sites to determine the cause or nature of disease.

Applications of Cytopathology

The main applications of cytology include the following:

- Screening for the early detection of asymptomatic cancer For example, the examination of scrapings from cervix for early detection and prevention of cervical cancer.
- Diagnosis of symptomatic cancer Cytopathology may be used alone or in conjunction with other modalities to diagnose tumors revealed by physical or radiological examinations.
- It can be used in the diagnosis of cysts, inflammatory conditions and infections of various organs.
- Surveillance of patients treated for cancer For some types of cancers, cytology is the most feasible method of surveillance to detect recurrence. The best example is periodic urine cytology to monitor the recurrence of cancer of the urinary tract.

Advantages of Cytologic Examination

Compared to histopathologic technique it is cheap, takes less time and needs no anesthesia to take specimens. Therefore, it is appropriate for developing countries with limited resources like Ethiopia. In addition, it is complementary to histopathological examination.

Cytopathologic Methods

There are different cytopathologic methods including:

Fine-needle Aspiration Cytology (FNAC)

In FNAC, cells are obtained by aspirating the diseased organ using a very thin needle under negative pressure. Virtually any organ or tissue can be sampled by fine-needle aspiration. The aspirated cells are then stained & are studied under the microscope. Superficial organs (e.g. thyroid, breast, lymph nodes, skin and soft tissues) can be easily aspirated. Deep organs, such as the lung, mediastinum, liver, pancreas, kidney, adrenal gland, and retroperitoneum are aspirated with guidance by fluoroscopy, ultrasound or CT scan. FNAC is cheap, fast, & accurate in diagnosing many diseases.

Exfoliative Cytology

Refers to the examination of cells that are shed spontaneously into body fluids or secretions. Examples include sputum, cerebrospinal fluid, urine, effusions in body cavities (pleura, pericardium, peritoneum), nipple discharge and vaginal discharge.

Abrasive Cytology

Refers to methods by which cells are dislodged by various tools from body surfaces (skin, mucous membranes, and serous membranes). e.g. preparation of cervical smears with a spatula or a small brush to detect cancer of the uterine cervix at early stages. Such cervical smears, also called Pap smears, can significantly reduce the mortality from cervical cancer. Cervical cancer is the most common cancer in Ethiopian women.

Hematological Examination

This is a method by which abnormalities of the cells of the blood and their precursors in the bone marrow are investigated to diagnose the different kinds of anemia & leukemia.

Immunohistochemistry

This is a method is used to detect a specific antigen in the tissue in order to identify the type of disease.

Microbiological Examination

This is a method by which body fluids, excised tissue, etc. are examined by microscopical, cultural and serological techniques to identify micro-organisms responsible for many diseases.

Biochemical Examination

This is a method by which the metabolic disturbances of disease are investigated by assay of various normal and abnormal compounds in the blood, urine, etc.

Clinical Genetics (Cytogenetics)

This is a method in which inherited chromosomal abnormalities in the germ cells or acquired chromosomal abnormalities in somatic cells are investigated using the techniques of molecular biology.

Molecular Techniques

Different molecular techniques such as fluorescent in situ hybridization, Southern blot, etc. can be used to detect genetic diseases.

Autopsy

Autopsy is examination of the dead body to identify the cause of death. This can be for forensic or clinical purposes.

- The relative importance of each of the above disciplines to our understanding of disease varies for different types of diseases. For example, in diabetes mellitus, biochemical investigation provides the best means of diagnosis and is of greatest value in the control of the disease. Whereas in the diagnosis of tumors, FNAC & histopathology contribute much. However, for most diseases, diagnosis is based on a combination of pathological investigations.

Causes of Disease

Diseases can be caused by either environmental factors, genetic factors or a combination of the two.

Environmental Factors

Environmental causes of disease are many and are classified into:

- Physical agents: These include trauma, radiation, extremes of temperature, and electric power. These agents apply excess physical energy, in any form, to the body.
- Chemicals: With the use of an ever-increasing number of chemical agents such as drugs, in industrial processes, and at home, chemically induced injury has become very common. Their effects vary:
    - Some act in a general manner, for example cyanide is toxic to all cells.
    - Others act locally at the site of application, for example strong acids and caustics.
    - Another group exhibit a predilection for certain organs, for example – the effect of paracetamol and alcohol on liver. Many toxic chemicals are metabolized in liver and excreted in kidney, as a result, these organs are susceptible to chemical injury.
- Nutritional deficiencies and excesses: Nutritional deficiencies may arise as a result of poor supply, interference with absorption, inefficient transport within the body, or defective utilization. It may take the form of deficiency either of major classes of food, usually protein and energy, or vitamins or elements essential for specific metabolic processes, e.g. iron for haemoglobin production. Often, the deficiencies are multiple and complex.

On the other hand, dietary excess plays an important role in diseases in Western countries. Obesity has become increasingly common, with its attendant dangers of type 2 diabetes, high blood pressure and heart disease.

- Infections and infestations: Viruses, bacteria, fungi, protozoa, and metazoa all cause diseases. They may do so by causing cell destruction directly as in virus infections (for example poliomyelitis) or protozoal infections (for example malaria). However, in others the damage is done by toxins elaborated by the infecting agent as in diphtheria and tetanus. Like chemicals, they may have a general effect or they may show a predilection for certain tissues.
- Immunological factors: The immune process is essential for protection against micro-organisms and parasites. However, the immune system can be abnormal which can lead to diseases. The abnormalities of the immune system include:
    - Hypersensitivity reaction: This is exaggerated immune response to an antigen. For example, bronchial asthma can occur due to exaggerated immune response to the harmless pollen.
    - Immunodeficiency: This is due to deficiency of a component of the immune system which leads to increased susceptibility to different diseases. An example is AIDS.
    - Autoimmunity: This is an abnormal (exaggerated) immune reaction against the self antigens of the host. Therefore, autoimmunity is a hypersensitivity reaction against the self antigens. For example, type 1 diabetes mellitus is caused by autoimmune destruction of the beta cells of the islets of Langerhans of the pancreas.
- Psychogenic factors: The mental stresses imposed by conditions of life, particularly in technologically advanced communities, are probably contributory factors in some groups of diseases.

Genetic Factors

These are hereditary factors that are inherited genetically from parents. Detailed discussion will be done on this topic in a subsequent chapter.

Course of Disease

The course of disease is shown with a simplified diagram as follows:

Exposure → Biological onset → Clinical onset → Permanent damage → Death

↑

Latency period

The course of a disease in the absence of any intervention is called the natural history of the disease. The different stages in the natural history of disease include:

- Exposure to various risk factors (causative agents).
- Latency, period between exposure and biological onset of disease.
- Biological onset of disease; this marks the initiation of the disease process, however, without any sign or symptom. Following biological onset of disease, it may remain asymptomatic or subclinical (i.e. without any clinical manifestations), or may lead to overt clinical disease.
- Incubation (induction) period refers to variable period of time without any obvious signs or symptoms from the time of exposure.
- The clinical onset of the disease, when the signs and symptoms of the disease become apparent. The expression of the disease may be variable in severity or in terms of range of manifestations.
- The onset of permanent damage.
- Death.

Natural recovery, i.e. recovery without any intervention, can occur at any stage in the progression of the disease.

Outcome and Consequences of Disease

Following clinical onset, disease may follow any of the following trends:

- Resolution can occur leaving no sequelae,
- The disease can settle down, but sequelae are left, or
- It may result in death.

Clinical and Biological Death

Clinical Death

Clinical death is the reversible transmission between life and biological death. Clinical death is defined as the period of respiratory, circulatory and brain arrest during which initiation of resuscitation can lead to recovery.

Clinical death begins with either the last agonal inhalation or the last cardiac contraction. Signs indicating clinical death are:

- The patient is without pulse or blood pressure and is completely unresponsive to the most painful stimulus.

- The pupils are widely dilated.
- Some reflex reactions to external stimulation are preserved. For example, during intubations, respiration may be restored in response to stimulation of the receptors of the superior laryngeal nerve, the nucleus of which is located in the medulla oblongata near the respiratory center.
- Recovery can occur with resuscitation.

Biological Death

Biological death (sure sign of death), which sets in after clinical death, is an irreversible state of cellular destruction. It manifests with irreversible cessation of circulatory and respiratory functions, or irreversible cessation of all functions of the entire brain, including brain stem. However, one should notice that there are internationally accepted criteria to diagnose biological death.

PATHOPHYSIOLOGY

Pathophysiology – a convergence of pathology with physiology – is the study of the disordered physiological processes that cause, result from, or are otherwise associated with a disease or injury. Pathology is the medical discipline that describes conditions typically *observed* during a disease state, whereas physiology is the biological discipline that describes processes or mechanisms *operating* within an organism. Pathology describes the abnormal or undesired condition, whereas pathophysiology seeks to explain the functional changes that are occurring within an individual due to a disease or pathologic state.

Reductionism

In Germany in the 1830s, Johannes Müller led the establishment of physiology research autonomous from medical research. In 1843, the Berlin Physical Society was founded in part to purge biology and medicine of vitalism, and in 1847 Hermann von Helmholtz, who joined the Society in 1845, published the paper "On the conservation of energy", highly influential to reduce physiology's research foundation to physical sciences. In the late 1850s, German anatomical pathologist Rudolf Virchow, a former student of Müller, directed focus to the cell, establishing cytology as the focus of physiological research, while Julius Cohnheim pioneered experimental pathology in medical schools' scientific laboratories.

Germ Theory

By 1863, motivated by Louis Pasteur's report on fermentation to butyric acid, fellow Frenchman Casimir Davaine identified a microorganism as the crucial causal agent of the cattle disease anthrax, but its routinely vanishing from blood left other scientists inferring it a mere byproduct of putrefaction. In 1876, upon Ferdinand Cohn's report of a tiny spore stage of a bacterial species, the fellow German Robert Koch isolated Davaine's *bacterides* in pure culture—a pivotal step that would establish bacteriology as a distinct discipline—identified a spore stage, applied Jakob Henle's postulates, and confirmed Davaine's conclusion, a major feat for experimental pathology. Pasteur and colleagues followed up with ecological investigations confirming its role in the natural environment via spores in soil.

Also, as to sepsis, Davaine had injected rabbits with a highly diluted, tiny amount of putrid blood, duplicated disease, and used the term *ferment of putrefaction*, but it was unclear whether this referred as did Pasteur's term *ferment* to a microorganism or, as it did for many others, to a chemical. In 1878, Koch published *Aetiology of Traumatic Infective Diseases*, unlike any previous work, where in 80 pages Koch, as noted by an historian, "was able to show, in a manner practically conclusive, that a number of diseases, differing clinically, anatomically, and in aetiology, can be produced experimentally by the injection of putrid materials into animals." Koch used bacteriology and the new staining methods with aniline dyes to identify particular microorganisms for each. Germ theory of disease crystallized the concept of cause—presumably identifiable by scientific investigation.

Scientific Medicine

The American physician William Welch trained in German pathology from 1876 to 1878, including under Cohnheim, and opened America's first scientific laboratory—a pathology laboratory—at Bellevue Hospital in New York City in 1878. Welch's course drew enrollment from students at other medical schools, which responded by opening their own pathology laboratories. Once appointed by Daniel Coit Gilman, upon advice by John Shaw Billings, as founding dean of the medical school of the newly forming Johns Hopkins University that Gilman, as its first president, was planning, Welch traveled again to Germany for training in Koch's bacteriology in 1883. Welch returned to America but moved to Baltimore, eager to overhaul American medicine, while blending Vichow's anatomical pathology, Cohnheim's experimental pathology, and Koch's bacteriology. Hopkins medical school, led by the "Four Horsemen"—Welch, William Osler, Howard Kelly, and William Halsted—opened at last in 1893 as America's first medical school devoted to teaching German scientific medicine, so called.

Biomedicine

The first biomedical institutes, Pasteur Institute and Berlin Institute for Infectious Diseases, whose first directors were Pasteur and Koch, were founded in 1888 and 1891, respectively. America's first biomedical institute, The Rockefeller Institute for Medical Research, was founded in 1901 with Welch, nicknamed "dean of American medicine", as its scientific director, who appointed his former Hopkins student Simon Flexner as director of pathology and bacteriology laboratories. By way of World War I and World War II, Rockefeller Institute became the globe's leader in biomedical research.

Molecular Paradigm

The 1918 pandemic triggered frenzied search for its cause, although most deaths were via lobar pneumonia, already attributed to pneumococcal invasion. In London, pathologist with the Ministry of Health, Fred Griffith in 1928 reported pneumococcal transformation from virulent to avirulent and between antigenic types—nearly a switch in species—challenging pneumonia's specific causation. The laboratory of Rockefeller Institute's Oswald Avery, America's leading pneumococcal expert, was so troubled by the report that they refused to attempt repetition.

When Avery was away on summer vacation, Martin Dawson, British-Canadian, convinced that anything from England must be correct, repeated Griffith's results, then achieved transformation *in vitro*, too, opening it to precise investigation. Having returned, Avery kept a photo of Griffith on his desk while his researchers followed the trail. In 1944, Avery, Colin MacLeod, and Maclyn McCarty reported the transformation factor as DNA, widely doubted amid estimations that something must act with it. At the time of Griffith's report, it was unrecognized that bacteria even had genes.

The first genetics, Mendelian genetics, began at 1900, yet inheritance of Mendelian traits was localized to chromosomes by 1903, thus chromosomal genetics. Biochemistry emerged in the same decade. In the 1940s, most scientists viewed the cell as a "sack of chemicals"—a membrane containing only loose molecules in chaotic motion—and the only especial cell structures as chromosomes, which bacteria lack as such. Chromosomal DNA was presumed too simple, so genes were sought in chromosomal proteins. Yet in 1953, American biologist James Watson, British physicist Francis Crick, and British chemist Rosalind Franklin inferred DNA's molecular structure—a double helix—and conjectured it to spell a code. In the early 1960s, Crick helped crack a genetic code in DNA, thus establishing molecular genetics.

In the late 1930s, Rockefeller Foundation had spearheaded and funded the molecular biology research program—seeking fundamental explanation of organisms and life—led largely by physicist Max Delbrück at Caltech and Vanderbilt University. Yet the reality of organelles in cells was controversial amid unclear visualization with conventional light microscopy. Around 1940, largely via cancer research at Rockefeller Institute, cell biology emerged as a new discipline filling the vast gap between cytology and biochemistry by applying new technology—ultracentrifuge and electron microscope—to identify and deconstruct cell structures, functions, and mechanisms. The two new sciences interlaced, *cell and molecular biology*.

Mindful of Griffith and Avery, Joshua Lederberg confirmed bacterial conjugation—reported decades earlier but controversial—and was awarded the 1958 Nobel Prize in Physiology or Medicine. At Cold Spring Harbor Laboratory in Long Island, New York,

Delbrück and Salvador Luria led the Phage Group—hosting Watson—discovering details of cell physiology by tracking changes to bacteria upon infection with their viruses, the process transduction. Lederberg led the opening of a genetics department at Stanford University's medical school, and facilitated greater communication between biologists and medical departments.

Disease Mechanisms

In the 1950s, researches on rheumatic fever, a complication of streptococcal infections, revealed it was mediated by the host's own immune response, stirring investigation by pathologist Lewis Thomas that led to identification of enzymes released by the innate immune cells macrophages and that degrade host tissue. In the late 1970s, as president of Memorial Sloan–Kettering Cancer Center, Thomas collaborated with Lederberg, soon to become president of Rockefeller University, to redirect the funding focus of the US National Institutes of Health toward basic research into the mechanisms operating during disease processes, which at the time medical scientists were all but wholly ignorant of, as biologists had scarcely taken interest in disease mechanisms. Thomas became for American basic researchers a patron saint.

Uses

Pathophysiology is a required area of study for nearly all healthcare professional school programs (medical, dental, physician assistant, occupational therapy, physical therapy, nurse practitioner, pharmacy, nursing, radiologic science and paramedic programs) in the United States, Canada and other countries.

Examples:

- The pathophysiology of Parkinson's disease is death of dopaminergic neurons as a result of changes in biological activity in the brain with respect to Parkinson's disease (PD). There are several proposed mechanisms for neuronal death in PD; however, not all of them are well understood. Five proposed major mechanisms for neuronal death in Parkinson's Disease include protein aggregation in Lewy bodies, disruption of autophagy, changes in cell metabolism or mitochondrial function, neuroinflammation, and blood-brain barrier (BBB) breakdown resulting in vascular leakiness.

- The pathophysiology of heart failure is a reduction in the efficiency of the heart muscle, through damage or overloading. As such, it can be caused by a wide number of conditions, including myocardial infarction (in which the heart muscle is starved of oxygen and dies), hypertension (which increases the force of contraction needed to pump blood) and amyloidosis (in which misfolded proteins are deposited in the heart muscle, causing it to stiffen). Over time these increases in workload will produce changes to the heart itself.

- The pathophysiology of multiple sclerosis is that of an inflammatory demyelinating disease of the CNS in which activated immune cells invade the central nervous system and cause inflammation, neurodegeneration and tissue damage. The underlying condition that produces this behaviour is currently unknown. Current research in neuropathology, neuroimmunology, neurobiology, and neuroimaging, together with clinical neurology provide support for the notion that MS is not a single disease but rather a spectrum
- The pathophysiology of hypertension is that of a chronic disease characterized by elevation of blood pressure. Hypertension can be classified by cause as either essential (also known as primary or idiopathic) or secondary. About 90–95% of hypertension is essential hypertension.
- The pathophysiology of HIV/AIDS involves, upon acquisition of the virus, that the virus replicates inside and kills T helper cells, which are required for almost all adaptive immune responses. There is an initial period of influenza-like illness, and then a latent, asymptomatic phase. When the CD4 lymphocyte count falls below 200 cells/ml of blood, the HIV host has progressed to AIDS, a condition characterized by deficiency in cell-mediated immunity and the resulting increased susceptibility to opportunistic infections and certain forms of cancer.
- The pathophysiology of spider bites is due to the effect of its venom. A spider envenomation occurs whenever a spider injects venom into the skin. Not all spider bites inject venom – a dry bite, and the amount of venom injected can vary based on the type of spider and the circumstances of the encounter. The mechanical injury from a spider bite is not a serious concern for humans.
- The pathophysiology of obesity involves many possible pathophysiological mechanisms involved in its development and maintenance. This field of research had been almost unapproached until the leptin gene was discovered in 1994 by J. M. Friedman's laboratory. These investigators postulated that leptin was a satiety factor. In the ob/ob mouse, mutations in the leptin gene resulted in the obese phenotype opening the possibility of leptin therapy for human obesity. However, soon thereafter J. F. Caro's laboratory could not detect any mutations in the leptin gene in humans with obesity. On the contrary Leptin expression was increased proposing the possibility of Leptin-resistance in human obesity.

PATHOPHYSIOLOGY OF HIV/AIDS

HIV is commonly transmitted via unprotected sexual activity, blood transfusions, hypodermic needles, and from mother to child. Upon acquisition of the virus, the virus replicates inside and kills T helper cells, which are required for almost all adaptive immune responses. There is an initial period of influenza-like illness, and then a latent, asymptomatic phase. When the CD4 lymphocyte count falls below 200 cells/ml of blood, the HIV host has progressed to AIDS, a condition characterized by deficiency in cell-mediated immunity and the resulting increased susceptibility to opportunistic infections and certain forms of cancer.

Immunology

After the virus enters the body there is a period rapid viral replication, leading to an abundance of virus in the peripheral blood. During primary infection, the level of HIV may reach several million virus particles per milliliter of blood.

This response is accompanied by a marked drop in the numbers of circulating CD4+ T cells. This acute viremia is associated in virtually all people with the activation of CD8+ T cells, which kill HIV-infected cells, and subsequently with antibody production, or seroconversion. The CD8+ T cell response is thought to be important in controlling virus levels, which peak and then decline, as the CD4+ T cell counts rebound. A good CD8+ T cell response has been linked to slower disease progression and a better prognosis, though it does not eliminate the virus.

During the acute phase, HIV-induced cell lysis and killing of infected cells by cytotoxic T cells accounts for CD4+ T cell depletion, although apoptosis may also be a factor. During the chronic phase, the consequences of generalized immune activation coupled with the gradual loss of the ability of the immune system to generate new T cells appear to account for the slow decline in CD4+ T cell numbers.

However, the symptoms of immune deficiency characteristic of AIDS do not appear for years after a person is infected, the bulk of CD4+ T cell loss occurs during the first weeks of infection, especially in the intestinal mucosa, which harbors the majority of the lymphocytes found in the body. The reason for the preferential loss of mucosal CD4+ T cells is that a majority of mucosal CD4+ T cells express the CCR5 coreceptor, whereas a small fraction of CD4+ T cells in the bloodstream do so.

HIV seeks out and destroys CCR5 expressing CD4+ cells during acute infection. A vigorous immune response eventually controls the infection and initiates the clinically latent phase. However, CD4+ T cells in mucosal tissues remain depleted throughout the infection, although enough remain to initially ward off life-threatening infections.

Continuous HIV replication results in a state of generalized immune activation persisting throughout the chronic phase. Immune activation, which is reflected by the increased activation state of immune cells and release of proinflammatory cytokines, results from the activity of several HIV gene products and the immune response to ongoing HIV replication. Another cause is the breakdown of the immune surveillance system of the mucosal barrier caused by the depletion of mucosal CD4+ T cells during the acute phase of disease.

This results in the systemic exposure of the immune system to microbial components of the gut's normal flora, which in a healthy person is kept in check by the mucosal immune system. The activation and proliferation of T cells that results from immune activation provides fresh targets for HIV infection. However, direct killing by HIV alone cannot account for the observed depletion of CD4+ T cells since only 0.01–0.10% of CD4+ T cells in the blood are infected.

A major cause of CD4+ T cell loss appears to result from their heightened susceptibility to apoptosis when the immune system remains activated. Although new T cells are continuously produced by the thymus to replace the ones lost, the regenerative capacity of the thymus is slowly destroyed by direct infection of its thymocytes by HIV. Eventually, the minimal number of CD4+ T cells necessary to maintain a sufficient immune response is lost, leading to AIDS.

CD4 T-cell Death and Inflammation

Recent studies employed an ex vivo human lymphoid aggregate culture (HLAC) system formed with fresh human tonsil or spleen tissue to model molecular and cellular events in human tissues during in vivo HIV infection. These studies found that >95% of CD4 T cells die because of abortive HIV infection. These dying cells are resting and thus are nonpermissive for productive HIV infection. Full viral replication was limited to the ~5% of activated CD4 T cells present in these tissues; these cells die by apoptosis. Abortive HIV infection occurs due to slowing of reverse transcription promoting cytosolic DNA accumulation. This viral DNA is sensed by gamma-interferon-inducible protein 16 (IFI16), which produces an innate immune response against the virus by activating caspase 1 in IFI16 inflammasomes and inducing pyroptosis, a highly inflammatory form of programmed cell death. These findings cast CD4 T-cell death during HIV infection in a different light. Rather than the virus playing a major role, it is the host response to viral DNA produced during abortive infection that triggers CD4 T-cell death. Further, these findings identify novel drug targets that may be exploited to both block CD4 T cell demise and the chronic inflammatory response generated during pyroptosis.

Cells Affected

The virus, entering through which ever route, acts primarily on the following cells:

- Lymphoreticular system:
  - CD4+ T-Helper cells (main target cell).
  - Macrophages.
  - Monocytes.
- Certain endothelial cells:

- Central nervous system:
    - Microglia of the nervous system.
    - Astrocytes.
    - Oligodendrocytes.
    - Neurones – indirectly by the action of cytokines and the gp-120.

Effect

Although the virus has cytopathic effects in productively infected cells, this effect may not directly contribute to HIV pathogenesis. Importantly, the virus can remain inactive (latent) in these productively infected cells for long periods.

- CD4 T-cell depletion and chronic inflammation are the two signature events that drive HIV pathogenesis and progression to AIDS.
- Infection of the cells of the CNS cause acute aseptic meningitis, subacute encephalitis, vacuolar myelopathy and peripheral neuropathy. Later it leads to even AIDS dementia complex.
- The CD_4-gp120 interaction is also permissive to other viruses like Cytomegalovirus, Hepatitis virus, Herpes simplex virus, etc. These viruses lead to further cell damage i. e. cytopathy.

PATHOPHYSIOLOGY OF HYPERTENSION

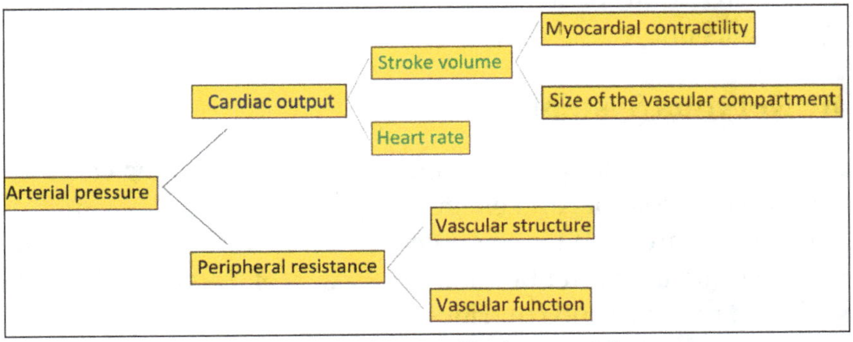

A diagram explaining factors affecting arterial pressure.

Pathophysiology is a branch of medicine which explains the function of the body as it relates to diseases and conditions. The pathophysiology of hypertension is an area which attempts to explain mechanistically the causes of hypertension, which is a chronic disease characterized by elevation of blood pressure. Hypertension can be classified by cause as either essential (also known as primary or idiopathic) or secondary. About 90–95% of hypertension is essential hypertension. Some authorities define essential hypertension as that which has no known explanation, while others define its cause as being due to overconsumption of sodium and underconsumption of potassium. Secondary hypertension indicates that the hypertension is a result of a specific underlying condition with a well-known mechanism, such as chronic kidney disease, narrowing of the aorta or kidney arteries, or endocrine disorders such as excess aldosterone, cortisol, or catecholamines. Persistent hypertension is a major risk factor for hypertensive heart disease, coronary artery disease, stroke, aortic aneurysm, peripheral artery disease, and chronic kidney disease.

Cardiac output and peripheral resistance are the two determinants of arterial pressure. Cardiac output is determined by stroke volume and heart rate; stroke volume is related to myocardial contractility and to the size of the vascular compartment. Peripheral resistance is determined by functional and anatomic changes in small arteries and arterioles.

Genetics

Single gene mutations can cause Mendelian forms of high blood pressure; ten genes have been identified which cause these monogenic forms of hypertension. These mutations affect blood pressure by altering kidney salt handling. There is greater similarity in blood pressure within families than between families, which indicates a form of inheritance, and this is not due to shared environmental factors. With the aid of genetic analysis techniques, a statistically significant linkage of blood pressure to several chromosomal regions, including regions linked to familial combined hyperlipidemia, was found. These findings suggest that there are many genetic loci, in the general population, each with small effects on blood pressure. Overall, however, identifiable single-gene causes of hypertension are uncommon, consistent with a multifactorial cause of essential hypertension.

Autonomic Nervous System

The autonomic nervous system plays a central role in maintaining cardiovascular homeostasis via pressure, volume, and chemoreceptor signals. It does this by regulating the peripheral vasculature, and kidney function, which in turn affect cardiac output, vascular resistance, and fluid retention. Excess activity of the sympathetic nervous system increases blood pressure and contributes to hypertension.

The mechanisms of increased sympathetic nervous system activity in hypertension involve alterations in baroreflex and chemoreflex pathways at both peripheral and central levels. Arterial baroreceptors are reset to a higher pressure in hypertensive patients, and this peripheral resetting reverts to normal when arterial pressure is normalized. Furthermore, there is central resetting of the aortic baroreflex in hypertensive patients, resulting in suppression of sympathetic inhibition after activation of aortic baroreceptor nerves. This baroreflex resetting seems to be mediated, at least partly, by a central action of angiotensin II. Additional small-molecule mediators that suppress baroreceptor activity and contribute to exaggerated sympathetic drive in hypertension include reactive oxygen species and endothelin. Some studies shown that hypertensive patients manifest greater vasoconstrictor responses to infused norepinephrine than normotensive controls. And that hypertensive patients do not show the normal response to increased circulating norepinephrine levels which generally induces downregulation of noradrenergic receptor, and it is believed that this abnormal response is genetically inherited.

Exposure to stress increases sympathetic outflow, and repeated stress-induced vasoconstriction may result in vascular hypertrophy, leading to progressive increases in peripheral resistance and blood pressure. This could partly explain the greater incidence of hypertension in lower socioeconomic groups, since they must endure greater levels of stress associated with daily living. Persons with a family history of hypertension manifest augmented vasoconstrictor and sympathetic responses to laboratory stressors, such as cold pressor testing and mental stress, that may predispose them to hypertension. This is particularly true of young African Americans. Exaggerated stress responses may contribute to the increased incidence of hypertension in this group.

Resistant hypertension can be treated by electrically stimulating the baroreflex with a pacemaker-like device.

Renin-angiotensin-aldosterone System

Another system maintaining the extracellular fluid volume, peripheral resistance, and that if disturbed may lead to hypertension, is the renin–angiotensin–aldosterone system. Renin is a circulating enzyme that participates in maintaining extracellular volume and arterial vasoconstriction, therefore contributing to regulation of blood pressure. It performs this function by breaking down (hydrolysing) angiotensinogen, secreted from the liver, into the peptide angiotensin I. Angiotensin I is further cleaved by an enzyme that is located primarily but not exclusively in the pulmonary circulation bound to endothelium; that enzyme is angiotensin converting enzyme (ACE). This cleavage produces angiotensin II, the most vasoactive peptide. Angiotensin II is a potent constrictor of all blood vessels. It acts on the musculature of arteries, raising peripheral resistance and thereby elevating blood pressure. Angiotensin II also causes the adrenal glands to release aldosterone, which stimulates the epithelial cells of the kidneys to increase re-absorption of salt and water, leading to raised blood volume and raised blood pressure. So elevated renin levels in the blood (normally 1.98-2.46 ng/ml in the upright position) leads to hypertension.

Recent studies claim that obesity is a risk factor for hypertension because of activation of the renin–angiotensin system (RAS) in adipose tissue, and also linked renin–angiotensin system with insulin resistance, and claims that anyone can cause the other.

Local production of angiotensin II in various tissues, including the blood vessels, heart, adrenals, and brain, is controlled by ACE and other enzymes, including the serine protease chymase. The activity of local renin–angiotensin systems and alternative pathways of angiotensin II formation may make an important contribution to remodeling of resistance vessels and the development of target organ damage (i.e. left ventricular hypertrophy, congestive heart failure, atherosclerosis, stroke, end-stage kidney disease, myocardial infarction, and arterial aneurysm) in hypertensive persons.

Endothelial Dysfunction

The endothelium of blood vessels produces an extensive range of substances that influence blood flow and, in turn, is affected by changes in the blood and the pressure of blood flow. For example, local nitric oxide and endothelin, which are secreted by the endothelium, are the major regulators of vascular tone and blood pressure. In patients with essential hypertension, the balance between the vasodilators and the vasoconstrictors is upset, which leads to changes in the endothelium and sets up a "vicious cycle" that contributes to the maintenance of high blood pressure. In patients with hypertension, endothelial activation and damage also lead to changes in vascular tone, vascular reactivity, and coagulation and fibrinolytic pathways. Alterations in endothelial function are a reliable indicator of target organ damage and atherosclerotic disease, as well as prognosis.

Evidence suggests that oxidant stress alters many functions of the endothelium, including modulation of vasomotor tone. Inactivation of nitric oxide (NO) by superoxide and other reactive oxygen species (ROS) seems to occur in conditions such as hypertension. Normally nitric oxide is an important regulator and mediator of numerous processes in the nervous, immune and cardiovascular systems, including smooth muscle relaxation thus resulting in vasodilation of the artery and increasing blood flow, suppressor of migration and proliferation of vascular smooth-muscle cells. It has been suggested that angiotensin II enhances formation of the oxidant superoxide at concentrations that affect blood pressure minimally.

Endothelin is a potent vasoactive peptide produced by endothelial cells that has both vasoconstrictor and vasodilator properties. Circulating endothelin levels are increased in some hypertensive patients, particularly African Americans and persons with hypertension.

Sodium/Potassium Ratio Hypothesis of Essential Hypertension

A 2007 review article states that while excessive sodium consumption has long been recognized as contributing to the risk of hypertension, "potassium, the main intracellular cation, has usually been viewed as a minor factor in the pathogenesis of hypertension. However, abundant evidence indicates that a potassium deficit has a critical role in hypertension and its cardiovascular sequelae." The authors state that modern, western, high sodium, low potassium diets result in corresponding changes in intracellular concentration of these, the two most important cations in animal cells. This imbalance leads to contraction of vascular smooth muscle, restricting blood flow and so driving up blood pressure. The authors cite studies which showing that potassium supplementation is effective in reducing hypertension.

Epidimiological support for this hypothesis can be found in a 2014 meta-analysis which states that "the sodium-to-potassium ratio appears to be more strongly associated with blood pressure outcomes than either sodium or potassium alone in hypertensive adult populations".

PATHOPHYSIOLOGY OF OBESITY

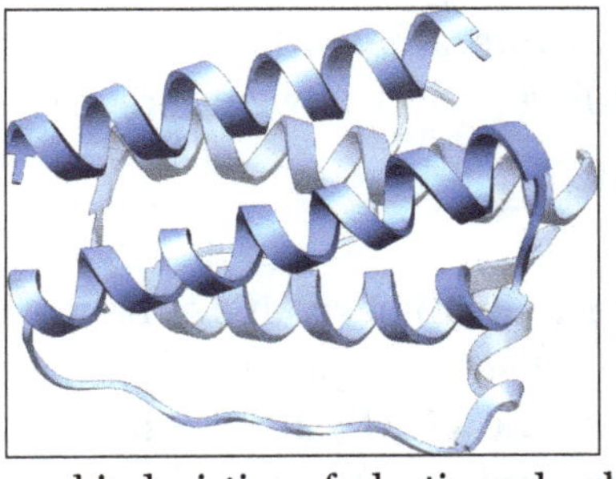

A graphic depiction of a leptin molecule.

There are many possible pathophysiological mechanisms involved in the development and maintenance of obesity.

This field of research had been almost unapproached until the leptin gene was discovered in 1994 by J. M. Friedman's laboratory. These investigators postulated that leptin was a satiety factor. In the ob/ob mouse, mutations in the leptin gene resulted in the obese phenotype opening the possibility of leptin therapy for human obesity. However, soon thereafter J. F. Caro's laboratory could not detect any mutations in the leptin gene in humans with obesity. On the contrary Leptin expression was increased proposing the possibility of Leptin-resistance in human obesity. Since this discovery, many other hormonal mechanisms have been elucidated that participate in the regulation of appetite and food intake, storage patterns of adipose tissue, and development of insulin resistance. Since leptin's discovery, ghrelin, insulin, orexin, PYY 3-36, cholecystokinin, adiponectin, as well as many other mediators have been studied. The adipokines are mediators produced by adipose tissue; their action is thought to modify many obesity-related diseases.

Appetite

Leptin and ghrelin are considered to be complementary in their influence on appetite, with ghrelin produced by the stomach modulating short-term appetitive control (i.e. to eat when the stomach is empty and to stop when the stomach is stretched). Leptin is produced by adipose tissue to signal fat storage reserves in the body, and mediates long-term appetitive controls (i.e. to eat more when fat storages are low and less when fat storages are high). Although administration of leptin may be effective in a small subset of obese individuals who are leptin-deficient, most obese individuals are thought to be leptin resistant and have been found to have high levels of leptin. This resistance is thought to explain in part why administration of leptin has not been shown to be effective in suppressing appetite in most obese people.

While leptin and ghrelin are produced peripherally, they control appetite through their actions on the central nervous system. In particular, they and other appetite-related hormones act on the hypothalamus, a region of the brain central to the regulation of food intake and energy expenditure. There are several circuits within the hypothalamus that contribute to its role in integrating appetite, the melanocortin pathway being the most well understood. The circuit begins with an area of the hypothalamus, the arcuate nucleus, that has outputs to the lateral hypothalamus (LH) and ventromedial hypothalamus (VMH), the brain's feeding and satiety centers, respectively.

Arcuate Nucleus

The arcuate nucleus contains two distinct groups of neurons. The first group coexpresses neuropeptide Y (NPY) and agouti-related peptide (AgRP) and has stimulatory inputs to the LH and inhibitory inputs to the VMH. The second group coexpresses pro-opiomelanocortin (POMC) and cocaine- and amphetamine-regulated transcript (CART) and has stimulatory inputs to the VMH and inhibitory inputs to the LH. Consequently, NPY/AgRP neurons stimulate feeding and inhibit satiety, while POMC/CART neurons stimulate satiety and inhibit feeding. Both groups of arcuate nucleus neurons are regulated in part by leptin. Leptin inhibits the NPY/AgRP group while stimulating the POMC/CART group. Thus a deficiency in leptin signaling, either via leptin deficiency or leptin resistance, leads to overfeeding and may account for some genetic and acquired forms of obesity.

Immune System

Obesity has been associated with an inflammatory state, which is chronic and low-grade inflammation, known as meta-inflammation. Meta-inflammation is subclinical meaning that while there is an increase in circulating pro-inflammatory factors, no clinical signs of inflammation, heat, pain, and redness, are seen with meta-inflammation. Based on the immune system cells involved, both innate and adaptive immunity are involved in meta-inflammation.

There are different types of obesity depending on where fat cells are stored. Abdominal obesity, excess fat cell accumulation in adipose tissue of the abdomen, is associated more strongly with meta-inflammation.

Evolutionarily, adipose tissue has been shown to function as an immune organ. The immune cells located in adipose tissue are important for maintaining metabolic homeostasis. With obesity, the immune cells important for maintaining metabolic homeostasis are suppressed because immune cell function and immune cell amount are affected by excess fat accumulation in adipose tissue. Excess fat accumulation can lead to insulin resistance, and insulin resistance has been linked to meta-inflammation. With insulin resistance, there is an increase in macrophages, mast cells, neutrophils, T lymphocytes, and B lymphocytes, and a decrease in eosinophils and some T lymphocytes.

Obesity has also been shown to induce hypoxic conditions in adipose cells. Hypoxic conditions result from fat cells expanding, and this decreases vascularization to surrounding fat cells. Decreased vascularization results in decreased amounts of oxygen in adipose tissue so adipose cells have to switch to anaerobic metabolism for energy production. Anaerobic metabolism stimulates inflammation caused by macrophages. There are two types of macrophages, classically activated M1 macrophages that increase inflammation and alternatively activated M2 macrophages that decrease inflammation. In animal studies, obesity has been shown to cause a shift from M2 to M1 macrophages in adipose tissue, causing an increase in inflammation.

PATHOPHYSIOLOGY OF MULTIPLE SCLEROSIS

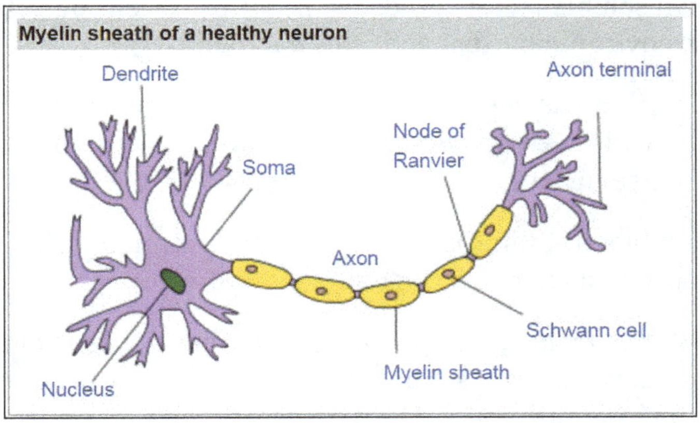

Multiple sclerosis is an inflammatory demyelinating disease of the CNS in which activated immune cells invade the central nervous system and cause inflammation, neurodegeneration, and tissue damage. The underlying cause is currently unknown. Current research in neuropathology, neuroimmunology, neurobiology, and neuroimaging, together with clinical neurology, provide support for the notion that MS is not a single disease but rather a spectrum.

There are three clinical phenotypes: Relapsing-remitting MS (RRMS), characterized by periods of neurological worsening following by remissions; secondary-progressive MS (SPMS), in which there is gradual progression of neurological dysfunction with fewer or no relapses; and primary-progressive MS (MS), in which neurological deterioration is observed from onset.

Pathophysiology is a convergence of pathology with physiology. Pathology is the medical discipline that describes conditions typically observed during a disease state; whereas physiology is the biological discipline that describes processes or mechanisms operating within an organism. Referring to MS, the physiology refers to the different processes that lead to the development of the lesions and the pathology refers to the condition associated with the lesions.

Pathology

Multiple sclerosis can be pathologically defined as the presence of distributed glial scars (or sclerosis) in the central nervous system disseminated in time (DIT) and space (DIS). The gold standard for MS diagnosis is pathological correlation, though given its limited availability, other diagnosis methods are normally used. The scleroses that define the disease are the remainders of previous demyelinating lesions in the CNS white matter of a patient (encephalomyelitis) showing special characteristics, such as confluent instead of perivenous demyelination.

There are two phases for how an unknown underlying condition may cause damage in MS:

- An unknown soluble factor (produced by CD8+ T-cells or CD20+ B-cells), creates a toxic environment that activates microglia.
- MRI-abnormal areas with hidden damage appear in the brain and spine (NAWM, NAGM, DAWM). Some clusters of activated microglia, axonal transection and myelin degeneration are present.
- Leaks in the blood–brain barrier appear and immune cells infiltrate, causing demyelination and axon destruction.

Multiple sclerosis differs from other idiopathic inflammatory demyelinating diseases in its confluent subpial cortical lesions. These types of lesions are the most specific finding for MS, being exclusively present in MS patients, though currently they can only be detected at autopsy.

Most MS findings take place inside the white matter, and lesions appear mainly in a periventricular distribution (clustered around the ventricles of the brain). Apart from white matter demyelination, the cortex and deep gray matter (GM) nuclei can be affected, together with diffuse injury of the NAWM. GM atrophy is independent of classical MS lesions and is associated with physical disability, fatigue, and cognitive impairment in MS.

At least five characteristics are present in CNS tissues of MS patients: Inflammation beyond classical white matter lesions, intrathecal Ig production with oligoclonal bands, an environment fostering immune cell persistence, and a disruption of the blood–brain barrier outside of active lesions. The scars that give the name to the condition are produced by astrocytes healing old lesions. MS is active even during remission periods.

Meningeal Tertiary Lymphoid-like Structures

Follicle-like aggregates in the meninges are formed only in secondary progressive MS. and correlate with the degree of subpial cortical demyelination and brain atrophy, suggesting that they might contribute to cortical pathology in SPMS.

These ectopic lymphoid follicles are composed mainly of EBV infected B-cells.

Physiology of MS

In multiple sclerosis, inflammation, demyelination, and neurodegeneration are observed together. Some clinical trials have shown that the inflammation produces both the relapses and the demyelination, and that neurodegeneration (axonal transection) is independent from inflammation, produces the accumulative disability, and advances even when inflammation and demyelination are delayed. It seems that neurodegeneration is produced by damaged mitochondria, which in turn comes from activated microglia.

Currently it is unknown what the primary cause of MS is; if MS is a heterogeneous disease, the lesion development process would not be unique. In particular, some PPMS patients having a special clinical course named rapidly progressive multiple sclerosis could have a special genetic cause and a different development process.

Several types of damage appear in the brain: *normal appearing white matter (NAWM)* and characteristic *lesions*. Changes in NAWM include axonal injury without demyelination, low-grade inflammation, and microglial and astrocytic activation

MS Lesion Development

MS lesions develop inside NAWM areas. Their shape is influenced by their activity.

The most accepted sequence of events is first NAWM appearance, then the so-called pre-active lesions, with activated microglia, and finally the BBB breakdown, which enables the entry of T-cells to the CNS. This marks the beginning of an autoimmune attack which destroys myelin in active lesions. When the attack is resolved, a characteristic glial scar is formed by astrocytes.

Current models can be divided into two categories: *inside-out* and *outside-in*. In the former, it is hypothesized that a problem in CNS cells produces an immune response that destroys myelin and subsequently breaks the BBB. In latter, an external factor produces BBB leaks, enters the CNS, and destroys myelin and axons. Whatever the underlying condition for MS is, it appears that damage is triggered by an unknown soluble factor in the CSF, potentially produced in meningeal areas; this factor can diffuse into the cortical parenchyma and destroy myelin either directly or indirectly through microglia activation.

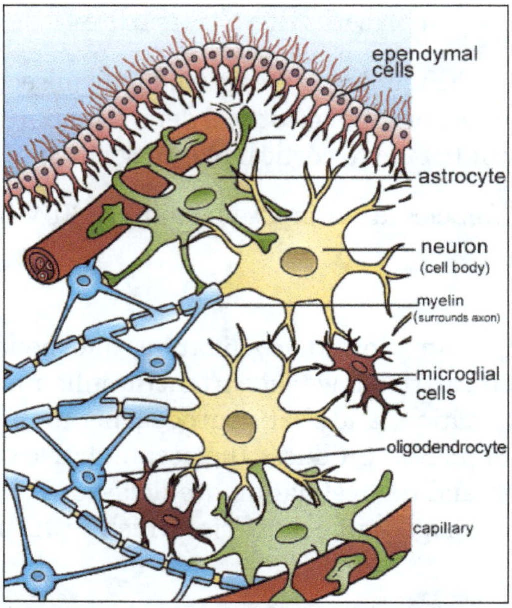

Illustration of the four different types of glial cells found in the central nervous system: ependymal cells, astrocytes, microglial cells, and oligodendrocytes.

The evolution of a preactive lesion is related to microglia reactivity. Increased expression of pro-inflammatory cell surface markers have been observed in NAWM and "initial" lesions, corresponding to a so-called loss of homeostatic microglial equilibrium.

Some authors report active lesion formation before BBB breakdown; others point to adipsin as a factor of the breakdown.

MS lesions are driven mainly by T-cells. It has been found recently that B-cells are also involved.

Blood–brain Barrier Disruption

The blood–brain barrier (*BBB*) is a protective barrier that denies the entrance of foreign material into the nervous system. BBB disruption is the moment in which penetration of the barrier by lymphocytes occur and has been considered one of the early problems in MS lesions.

The BBB is composed of endothelial cells which line the blood vessel walls of the central nervous system. Compared to normal endothelial cells, the cells lining the BBB are connected by occludin and claudin which form tight junctions in order to create a barrier to keep out larger molecules such as proteins. In order to pass through, molecules must be taken in by transport proteins or an alteration in the BBB permeability must occur, such as interactions with associated adaptor proteins like ZO-1, ZO-2 and ZO-3.

The BBB is compromised due to active recruitment of lymphocytes and monocytes and their migration across the barrier. Release of chemokines allow for the activation of adhesion molecules on the lymphocytes and monocytes, resulting in an interaction with the endothelial cells of the BBB which then activate the expression of matrix metalloproteinases to degrade the barrier. This results in disruption of the BBB, causing an increase in barrier permeability due to the degradation of tight junctions which maintain barrier integrity. Inducing the formation of tight junctions can restore BBB integrity and reduces its permeability, which can be used to reduce the damage caused by lymphocyte and monocyte migration across the barrier as restored integrity would restrict their movement.

After barrier breakdown symptoms may appear, such as swelling. Activation of macrophages and lymphocytes and their migration across the barrier may result in direct attacks on myelin sheaths within the central nervous system, leading to the characteristic demyelination event observed in MS. After demyelination has occurred, the degraded myelin sheath components, such as myelin basic proteins and Myelin oligodendrocyte glycoproteins, are then used as identifying factors to facilitate further immune activity upon myelin sheaths. Further activation of cytokines is also induced by macrophage and lymphocyte activity, promoting inflammatory activity as well continued activation of proteins such as matrix metalloproteinases, which have detrimental effect on BBB integrity.

Recently it has been found that BBB damage happens even in non-enhancing lesions. MS has an important vascular component.

Postmortem BBB Study

Postmortem studies of the BBB, especially the vascular endothelium, show immunological abnormalities. Microvessels in periplaque areas coexpressed HLA-DR and VCAM-1, some others HLA-DR and urokinase plasminogen activator receptor, and others HLA-DR and ICAM-1.

In Vivo BBB

The damaged white matter is known as "Normal-appearing white matter" (NAWM) and is where lesions appear. These lesions form in NAWM before blood–brain barrier breakdown.

BBB can be broken centripetally (the most normal) or centrifugally. Several possible biochemical disrupters were proposed. Some hypotheses about how the BBB is compromised revolve around the presence of compounds in the blood that could interact with vessels only in the NAWM areas. The permeability of two cytokines, Interleukin 15 and LPS, may be involved in BBB breakdown. Breakdown is responsible for monocyte infiltration and inflammation in the brain. Monocyte migration and LFA-1-mediated attachment to brain microvascular endothelia is regulated by SDF-1alpha through Lyn kinase.

Using iron nanoparticles, involvement of macrophages in BBB breakdown can be detected. A special role is played by Matrix metalloproteinases. These increase BBB T-cell permeability, specially in the case of MMP-9 and are supposedly related to the mechanism of action of interferons.

Whether BBB dysfunction is the cause or the consequence of MS is disputed, because activated T-Cells can cross a healthy BBB when they express adhesion proteins. Apart from that, activated T-Cells can cross a healthy BBB when they express adhesion proteins. (Adhesion molecules could also play a role in inflammation) In particular, one of these adhesion proteins involved is ALCAM (Activated Leukocyte Cell Adhesion Molecule, also called CD166), and is under study as therapeutic target. Another protein involved is CXCL12, which is found also in brain biopsies of inflammatory elements, and which could be related to the behavior of CXCL13 under methylprednisolone therapy. Some molecular biochemical models for relapses have been proposed.

Normally, gadolinium enhancement is used to show BBB disruption on MRIs. Abnormal tight junctions are present in both SPMS and PPMS. They appear in active white matter lesions and in gray matter in SPMS. They persist in inactive lesions, particularly in PPMS.

A uric acid deficiency was implicated in this process. Uric acid added in physiological concentrations (i.e. achieving normal concentrations) is therapeutic in MS by preventing BBB breakdown through inactivation of peroxynitrite. The low level of uric acid found in MS victims is manifestedly causative rather than a tissue damage consequence in the white matter lesions, but not in the grey matter lesions. Uric acid levels are lower during relapses.

Proposed Causes

It is not known what causes MS. Several problems appear together with the white matter lesions, like cortical lesions and normal-appearing tissues. Several theories have been proposed to explain it.

Some areas that appear normal under normal MRI look abnormal under special MRI, like magnetisation transfer MTR-MRI. These are called Normal Appearing White Matter (NAWM) and Normal Appearing Grey Matter (NAGM). The cause why the normal appearing areas appear in the brain is unknown, but seems clear that they appear mainly in the ventricles and that they predict the course of the disease.

Given that MS lesions begin inside the NAWM areas, these areas are expected to be produced by the same underlying condition that produces the lesions, and therefore the ultimate MS underlying condition, whatever it is. Historically, several theories about how these areas appear have been presented.

Autoimmune Theories

The search for an auto-antigen has taken a long time, but at least there is one reported. It is the enzyme GDP-L-fucose synthase.

This theory in part could also explain why some patients report amelioration under dietary treatment.

HERVs

Human endogenous retroviruses (HERVs) have been reported in MS for several years. In fact, one of the families, Human Endogenous Retrovirus-W was first discovered while studying MS patients.

Recent research as of 2019 point to one of the HERV-W viruses (pHEV-W), and specifically one of the proteins of the viral capside that has been found to activate microglia in vitro. Activated microglia in turn produces demyelination. Some interactions between the Epstein-Barr virus and the HERVs could be the trigger of the MS microglia reactions. Supporting this study, a monoclonal antibody against the viral capside (Temelimab) has shown good results in trials in phase IIb.

Blood Flow Related Theories

- Venous pathology has been associated with MS for more than a century. Pathologist Georg Eduard Rindfleisch noted in 1863 that the inflammation-associated lesions were distributed around veins. Some other authors like Tracy Putnam pointed to venous obstructions.
- Mechanical flow: Later the focus moved to softer hemodynamic abnormalities, which were showing precede changes in sub-cortical gray matter and in substantia nigra. However, such reports of a "hemodynamic cause of MS" are not universal, and possibly not even common. At this time the evidence is largely anecdotal and some MS patients have no blood flow issues. Possibly vascular problems may be an aggravating factor, like many others in MS. Indeed, the research, by demonstrating patients with no hemodynamic problems actually prove that this is not the only cause of MS.
- Endothelium: Other theories point to a possible primary endothelial dysfunction. The importance of vascular misbehaviour in MS pathogenesis has also been independently confirmed by seven-tesla MRI. It is reported that a number of studies have provided evidence of vascular occlusion in MS, which suggest the possibility of a primary vascular injury in MS lesions or at least that they are occasionally correlated.
- Venous insufficiency: Some morphologically special medullar lesions (wedge-shaped) have also been linked to venous insufficiency.

- BBB infection: It has also been pointed out that some infectious agents with positive correlation to MS, specially Chlamydia pneumoniae, can cause problems in veins and arteries walls.
- CCSVI: The term "chronic cerebrospinal venous insufficiency" was coined in 2008 by Paolo Zamboni, who described it in patients with multiple sclerosis. Instead of intracranial venous problems he described extracranial blockages, and he stated that the location of those obstructions seemed to influence the clinical course of the disease. According to Zamboni, CCSVI had a high sensitivity and specificity differentiating healthy individuals from those with multiple sclerosis. Zamboni's results were criticized as some of his studies were not blinded and they need to be verified by further studies. As of 2010 the theory is considered at least defensible.
- A more detailed evidence of a correlation between the place and type of venous malformations imaged and the reported symptoms of multiple sclerosis in the same patients was published in 2010.
- Haemodynamic problems have been found in the blood flow of MS patients using Doppler, initially using transcranial color-coded duplex sonography (TCCS), pointing to a relationship with a vascular disease called chronic cerebro-spinal venous insufficiency (CCSVI). In 2010 there were conflicting results when evaluating the relationship between MS and CCSVI. but is important to note that positives have appeared among the blinded studies.
- CSF flow: Other theories focus in the possible role of cerebrospinal fluid flow impairment. This theory could be partially consistent with the previous one. Currently a small trial with 8 participants has been performed.

CSF Composition: Kir4.1 and Anoctamin-2

Whatever the underlying primary condition is, it is expected to be a soluble factor in the CSF, maybe an unknown cytokine or ceramide, or a combination of them. Also B-cells and microglia could be involved. In particular, it is known that B-cells of MS patients secrete an unknown toxin against oligodendrocytes.

It has been reported several times that CSF of some MS patients can damage myelin in culture and mice and ceramides have been recently brought into the stage. Whatever the problem is, it produces apoptosis of neurons respecting astrocytes.

In 2012 it was reported that a subset of MS patients have a seropositive anti-Kir4.1 status, which can represent up to a 47% of the MS cases, and the study has been reproduced by at least two other groups.

In 2016 a similar association was reported for anti-Anoctamin-2.

If the existence of any of these subsets of MS is confirmed, the situation would be similar to what happened for Devic Disease and Aquaporin-4. MS could be considered a heterogeneous condition or a new medical entity will be defined for these cases.

Primary Neuro-degeneration Theories

Some authors propose a primary neurodegenerative factor. Maybe the strongest argument supporting this theory comes from the comparison with NMO. Though autoimmune demyelination is strong, axons are preserved, showing that the standard model of a primary demyelination cannot be hold. The theory of the trans-synaptic degeneration, is compatible with other models based in the CSF biochemistry.

Others propose an oligodendrocyte stress as primary dysfunction, which activates microglia creating the NAWM areas and others propose a yet-unknown intrinsic CNS trigger induces the microglial activation and clustering, which they point out could be again axonal injury or oligodendrocyte stress.

Finally, other authors point to a cortical pathology which starts in the brain external layer (pial surface) and progresses extending into the brain inner layers.

Genetic Causes

If as expected MS is an heterogeneus disease and the lesion development process would not be unique. In particular, some PPMS patients have been found to have a special genetic variant named rapidly progressive multiple sclerosis which would behave differently from what here is explained.

It is due to a mutation inside the gene NR1H3, an arginine to glutamine mutation in the position p.Arg415Gln, in an area that codifies the protein LXRA.

Biomarkers

Several biomarkers for diagnosis, disease evolution and response to medication (current or expected) are under research. While most of them are still under research, there are some of them already well stablished:

- Oligoclonal bands: They present proteins that are in the CNS or in blood. Those that are in CNS but not in blood suggest a diagnosis of MS.
- *MRZ-Reaction:* A polyspecific antiviral immune response against the viruses of measles, rubella and zoster found in 1992. In some reports the MRZR showed a lower sensitivity than OCB (70% vs. 100%), but a higher specificity (69% vs. 92%) for MS.
- Free light chains (FLC): Several authors have reported that they are comparable or even better than oligoclonal bands.

IMMUNO-PATHOPHYSIOLOGY OF MULTIPLE SCLEROSIS

Pathologic Characterization of MS Lesions

Plaques of inflammatory demyelination within the CNS are the pathologic hallmark of MS. Myelin destruction is an essential element of the plaque. Yet the MS plaque is not simply a static entity of myelin loss in isolation; rather, lesions are comprised of a wide variation of immunologic and pathologic features. These features have been categorized in an effort to understand the neural-immune mechanisms underlying MS. Constructing a framework around which myelin loss and neuronal/axonal injury occur entails a close examination of the cellular and molecular constituents, timing of damage, and repair processes. Thus, the various features of plaques provide a platform on which hypotheses regarding pathogenic mechanisms underlying MS have been formulated for over a century.

Traditionally, MS plaque classification has been based on temporal progression, or stages, of inflammatory destruction. Accordingly, acute, chronic active, and chronic silent lesions are thought to occur along a continuous timeline, eventually producing the scarred and hardened areas within the CNS that can be appreciated grossly. The acute MS plaque represents the earliest stage of lesion formation. It is typified by robust inflammatory infiltration combined with demyelination distributed throughout the lesion. Typical features of the acute plaque include ill-defined margins of myelin loss, infiltration of immune cells and parenchymal edema. The constituents of immune cell influx centered around vessels (termed perivascular cuffing) include lymphocytes (predominantly T cells), monocytes and macrophages. A portion of major histocompatibility class II (MHCII)-expressing cells, distributed evenly throughout the lesion, are loaded with lipids and participate in active stripping of myelin from axons. While oligodendrocyte apoptosis has been observed, the degree of oligodendrocyte loss within active lesions can be variable. In spite of the relative degree of axonal sparing, axonal injury can be extensive in acute lesions. Glial reactivity throughout the lesion is noted, particularly hypertrophic astrocytes. However, dense glial scarring is not typical of the acute plaque.

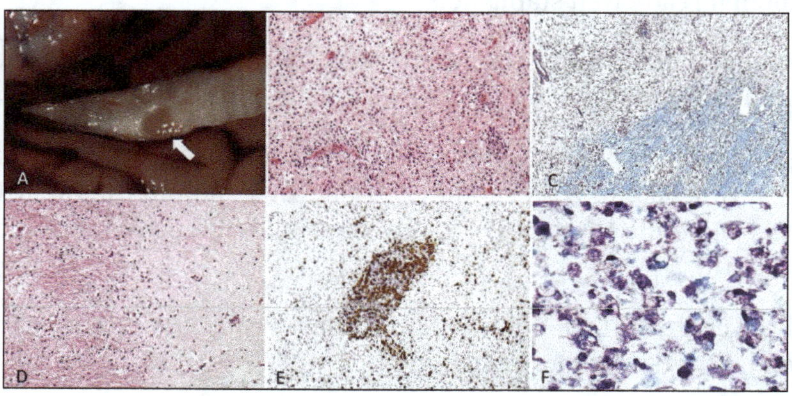

Gross and histologic features of MS plaques. (A) Gross examination of the brain at autopsy of a 79 year old with RR-MS. A dorsal view of the corpus callosum after separation of the cerebral hemispheres reveals a hardened, discolored area within the body of the corpus callosum (arrow). (B) Acute MS plaque stained with Hematoxylin and Eosin revealing hypercellularity due to the perivascular and parenchymal infiltration of leukocytes. (C) Luxol fast blue/periodic acid–Schiff (LFB-PAS) stained section of a plaque margin reveals a blurred but discrete edge (arrows). (D) Inactive plaque demonstrating borders that are distinct and devoid of inflammation (E) CD^{3+} lymphocytes clustering in a perivascular cuff within an active lesion area (F) Foamy macrophages characterized by fragments of myelin (arrow) engulfed by macrophages are identified using LFB-PAS at a plaque margin.

The chronic plaque is characterized by a region of hypocellularity with loss of myelin and glial scarring. On gross examination of post-mortem tissue, the hardened and discolored appearance of chronic plaques is often appreciable in frequently targeted areas of the CNS. Histologically, the lesion borders of chronic plaques are more distinct than those of acute plaques. A division into two forms of chronic plaque is made to signify temporal evolution from active destruction at the edge of the lesion (chronic active plaque) to an entirely "burned out" lesion devoid of active inflammatory destruction (chronic silent plaque). In chronic active lesions, inflammation continues along the outer border with the histologic appearance comparable to acute lesions. Thus, borders of the chronic active plaque are populated with activated microglia and macrophages, vessels demonstrating perivascular cuffing, and reactive astrocytes. The presence of antibody and complement are more prominent in chronic active lesions. Areas of re-myelination are often observed on the edge of lesions, but can encompass the entire of lesion. The core of the chronic plaque is typically hypocellular, though, and often contains thickened vessels with enlarged perivascular spaces. Chronic silent lesions are characterized by loss of the inflammatory traits along the border of chronic active lesions. Remyelination and the presence of oligodendrocyte progenitors are uncommon. Essentially, the chronic silent plaque is "burned out", containing minimal inflammation and devoid of the active inflammatory border. This is accompanied by a complete loss of oligodendrocytes and a variable but demonstrable reduction in axonal density. Overall, the gross and histologic features of the MS plaque imply a complex progression of inflammatory damage culminating in a scarred region of demyelination.

Extensive variability across MS plaques exists, including different degrees of inflammation, demyelination, re-myelination and axonal injury. Controversy has emerged in the "taxonomy" by which lesions are grouped, possibly in part as a result of studying a combination of autopsy and biopsy tissues at different stages of the disease. A relatively recent attempt was made to re-define the process of categorizing MS lesions by proposing a system for classification based on suspected pathologic mechanisms. Alongside a temporally-based system, Lucchinetti et al described various types of lesions based on the pattern of leukocyte markers, myelin proteins, immunoglobulin and complement within lesions. From 83 pathologic specimens obtained at autopsy or from biopsy, four distinct patterns of immunologic and pathologic features of actively demyelinating lesions were discerned, patterns II and III being the most common.

Pattern I is characterized by the predominance of T cell and macrophage inflammatory content; pattern II is characterized by T cell and macrophage infiltration, along with accentuated immunoglobulin deposition and myelin degradation products within macrophages; pattern III is distinguished by pronounced oligodendrocyte loss at the active edge of the lesion and preferential loss of myelin-associated glycoprotein (MAG); and pattern IV is characterized by oligodendrocyte dystrophy and the absence of remyelination or shadow plaques. Importantly, these distinct patterns were consistently observed within samples from the same patient, but varied between patients. The authors, based on the results from one open-label clinical trial in patients who had brain biopsies, hypothesized that deficit recovery after plasma exchange was indicative of the individual pattern of MS pathology. Plasma exchange was effective in patients with biopsy-proven pattern II (with prominent antibody and complement involvement) but not for patients with pattern I or III lesions. Overall, categorization of MS lesions based on immuno-pathologic mechanisms has added to the complexity of paradigms for lesion classification.

While the thought that distinct pathologic processes underlie a singular clinical entity referred to as MS is attractive, this concept has recently been disputed. In particular, challenges have emerged to the idea that there is a clear delineation between the different patterns of pathology. Prineas and colleagues have reported acute lesions from the same patient consisting of features from several of the four pathologic patterns. Furthermore, in a recent study, 131 tissue specimens from 39 MS patients all had lesions with little inter-individual variation in pathologic features. Indeed, lesions from this study exhibited traits consistent with inclusion into pattern II (i.e. deposition of complement and antibodies in proximity to macrophages and the absence of oligodendrocyte apoptosis). Therefore, various pathologic patterns of disease may more precisely refer to the stage of lesion; the relevance of the classification system based on four individual subtypes of plaques remains to be confirmed.

One hypothesis proposed regarding initial events during plaque genesis is that early intrinsic oligodendrocyte injury leads to the inflammatory damage historically associated with the pathologic features of MS plaques. Thus, toxic insults directly affecting the oligodendrocyte could serve as a trigger for a common immuno-pathologic pathway associated with evolution into later stages of the plaque. That 30 out of 39 cases examined by Breij et al were from patients who were in the progressive phase of disease would suggest that early heterogeneity of lesions could have been missed. Further support of this hypothesis comes from a study in which 26 active lesions were examined from 15 patients at autopsy, 11 of whom were early in the course of their disease. In this study, tissue immediately adjacent to lesion borders showed microscopic evidence of cellular injury without the presence of immune infiltration. Prineas and colleagues speculate that there are possibly toxic factors diffusible to the edge of the lesion resulting in oligodendrocyte fragility, and refer to these areas of initial injury as "pre-phagocytic".

Therefore, a critical question remaining in the field of neuropathology is whether differences exist between borders of acute versus chronic active lesions. Further work is required in order to more clearly identify white matter within the CNS that is undergoing early changes or at risk for doing so and how this relates to features described in more established MS plaques.

Normal Appearing White Matter

There has been a long-standing interest in early white matter changes in MS, with some of the earliest studies on myelin examining tissue outside of plaque regions that appears grossly unaffected. More recently, investigation into normal appearing white matter (NAWM) defined on conventional MRI sequences has been explored extensively using a variety of novel neuro-imaging techniques that show abnormalities in these areas suggestive of decreased myelin integrity and diminished axonal density within non-lesional regions.

Histo-pathologic examination of NAWM in patients with MS also support the concept that areas outside of plaques have immuno-pathologic changes. Microglial activation, T cell infiltration and perivascular cuffing have been reported in NAWM. Not surprisingly, these features were found more diffusely in cases of progressive MS. As a transition zone, dirty appearing white matter (DAWM) - defined by MRI as having intermediate signal intensity between NAWM and T2 lesions - showed diffuse loss of myelin histologically. Thus, greater involvement of NAWM in MS, associated with the progression of disease, may be related to the extent to which inflammation extends beyond focal lesions. Overall, the extent of white matter abnormalities and the immuno-pathologic features of NAWM have yet to be precisely defined.

Gray Matter Plaques

MS plaques are not confined to the white matter. Gray matter lesions are detected by MRI and by examination of pathologic specimens. Almost all of the gray matter nuclei within the CNS can be affected, as observed in a cohort of mostly progressive MS patients, but out of several regions of the CNS, including motor cortex, the spinal cord and cerebellum are particularly vulnerable, resulting in demyelination in up to 28.8% of the gray matter on average. As might be expected, inflammatory lesions within the gray matter are associated with neuronal loss and transected axons, which are more common in active lesions. While gray matter lesions seem to be more common in patients with progressive forms of MS, they can develop early in the disease process and possibly account for some of the disability seen in MS patients that cannot be explained by white matter lesions. For example, lesions located within the gray matter correlate better with cognitive disability than white matter lesions.

There are several unique features of pathology associated with gray matter lesions. Histologically, gray matter lesions are less inflammatory with fewer infiltrating T-lymphocytes and microglia/macrophages. Purely cortical gray matter lesions have also been described as lacking complement deposition and blood-brain barrier breakdown. Whether gray matter plaques arise from distinct immunologic mechanisms is unclear at this time. In an attempt to more carefully evaluate the processes involved, cortical lesions have been separated into several categories based on the depth of penetration from the surface into the brain. Type I lesions include discernable injury to both white and gray matter; Type II lesions have perivascular areas of demyelination isolated to the cortex; and Type III lesions demonstrate cortical demyelination below the pial surface that often cover several gyri and stop at cortical layers three or four. Others have proposed a fourth category, Type IV, to describe lesions that affect all cortical layers without extending into the white matter. Type III and IV lesions are the most extensive and difficult to visualize. Although scarce inflammatory cells are found within these lesions, the meninges overlying them contain inflammatory cells that collect in structures resembling ectopic B-cell follicles. In support of a role for lymphoid neo-genesis in the pathogenesis of these lesions is that patients with ectopic B-cell follicles (41% of secondary progressive MS patients) had a more rapid disease progression. More solid determination of the role for ectopic B cell follicle formation in MS will depend upon the reliable detection of cortical lesions by MRI and identification of the immuno-pathologic mechanisms leading to lymphoid neogenesis within the CNS.

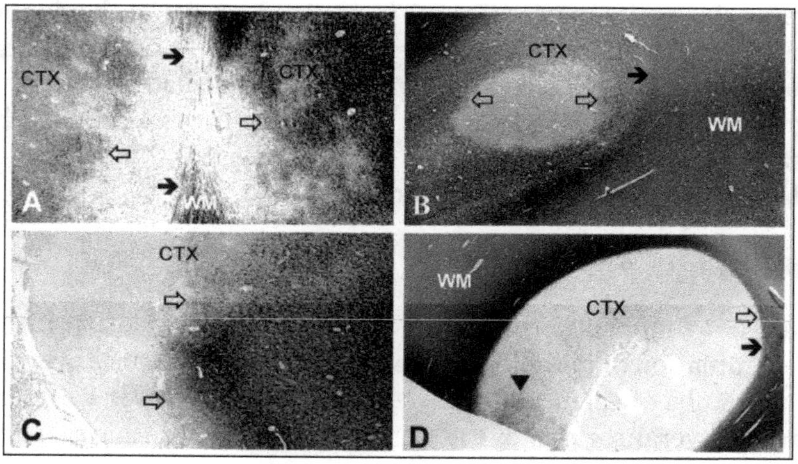

Cortical lesions in MS revealed by sections immunohistochemically labeled for MBP. (A) Type 1 cortical lesions encompass both white and gray matter. Solid arrows indicate the cortex (CTX)/white matter (WM) border, while the lesion border is delineated by open arrows. (B) Type 2 lesions are contained entirely within the CTX and do not extend to the subcortical WM (solid arrow) or pial surface. The border of this type 2 lesion is indicated by the open arrows. (C) Type III cortical lesions represent cortical demyelination below the pial surface that often cover several gyri and stop at cortical layers three or four. Open arrows indicate the lesion border. (D) Type 4 lesions (border represented by filled arrow) span the entirety of the CTX without involvement of WM (solid arrow). The arrowhead indicates a small area likely undergoing remyelination. Figure from used with permission.

Role of T Lymphocytes in Multiple Sclerosis

The lymphocytic presence within plaques and bordering areas suggests that inflammatory destruction in MS is driven by antigen-specific targeting of myelin and other CNS components. In particular, adaptive immune responses by T lymphocytes are thought to mediate injury to myelin and nerves within the CNS during MS. The determination that EAE can be mediated by CD4 T cells has promoted intense investigation into the potential CD4 T cell targets in MS. The relevance of antigenspecific CD4 T cell responses in MS was highlighted by the results of trials using an altered peptide ligand of MBP designed for therapeutic suppression of CD4 T cell responses, which resulted in disease exacerbations in multiple patients. T cells from MS patients can recognize a variety of myelin protein targets, including MBP, PLP, MOG and MOBP, among others. Non-myelin T cell antigens have also been described, including αB crystalline and neuronal proteins such as contactin-2. Further, auto-reactive CD8 T cell are also observed. Although similar frequencies of auto-reactive T cells are found in MS patients and healthy subjects, myelin-specific T cell avidity and activation profiles appear to be elevated in MS patients.

Newer technologies such as arrays of protein , lipid and gene expression have allowed fresh insight into targets that were heretofore unknown. For example, highly expressed genes in MS plaques compared to non-lesional CNS tissue include osteopontin ; however, elevated gene expression does not necessarily imply targeting by the adaptive immune response in MS. Genes encoding myelin proteins actually display reduced expression. Thus, while several of these targets have been validated in EAE, their involvement in MS has yet to be fully demonstrated. Again, it is likely that multiple CNS antigens exist for T cells during the inflammatory targeting in MS; results from array experiments highlight how diverse and extensive these targets may be.

How these T cells become abnormally activated toward CNS antigens remains unclear. A popular concept invoked to explain how T cells become pathogenic in MS patients is that of molecular mimicry. Several infectious agents have been postulated to serve as activation triggers for auto-reactive T cells. The most consistently reported one is Epstein Barr virus (EBV). Other infectious agents may also serve to trigger cross-reactivity with myelin components. One advantage of the adaptive immune response in MS lies in the ability to harness the specificity of T cell responses. As such, very discrete immunologic targets of therapy may be available for treating MS, including vaccines and antigen presenting cell (APC)-mediated delivery of antigens designed to induce antigenspecific tolerance. Clearly, antigen-specific therapeutics for MS must address the potentially wide array of T cell targets as well as HLA-specific presentation of antigens intrinsic to the genetic heterogeneity of individuals with MS.

Role of Th17 Cells in MS

Since the identification of IL-17 as a novel cytokine in 1993, an intense inquiry into the role of IL-17 in EAE and MS has taken place. Two members of the IL-17 gene family, IL-17A and IL-17F, are expressed by CD4 T cells. In 2003, Cua and colleagues identified IL-23, and not IL-12, as a critical cytokine regulator of EAE. This quickly led to a new pathway for investigation into the immunologic basis for MS after the discovery that IL-23 regulates IL-17 production by CD4 T cells. Subsequently, IL-17-secreting CD4 T cells have emerged as a distinct lineage of T helper cells, termed Th17 cells. In EAE, Th17 cells participate in early infiltration of the CNS and alone induce a unique neutrophil-predominant pathology. However, unlike IL-23, neutralization or genetic deletion of IL-17 in EAE does not result in a complete absence of disease. Hence, the direct contribution of Th17 cells during autoimmune demyelination remains unclear.

Human studies have emerged to bolster the relevance of IL-17 to the pathogenesis of MS. A greater proportion of IL-17-secreting cells, but not IFNγ+ CD4 T cells, is found in the CSF of MS patients compared to those with non-inflammatory neurologic diseases. The percentage of IL-17-producing memory CD4 T cells is elevated in peripheral blood from MS patients experiencing relapses, suggesting a prominent role for Th17 cells in MS. Further, IL-17 gene expression is up-regulated in lesions of MS patients, and Th17 cells are found in perivascular cuffs and borders of active lesions, indicating Th17 cells home to areas of inflammatory demyelination. These studies clearly highlight the association between Th17 cells and MS immune pathology.

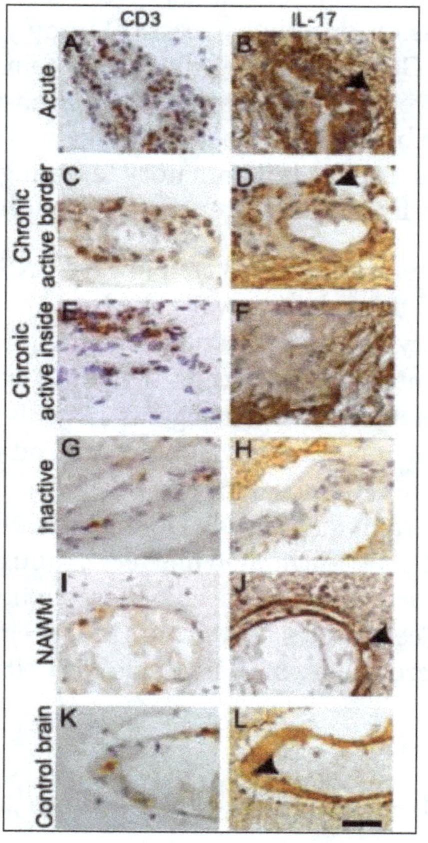

Involvement of IL-17 in MS lesions. Immunohistochemical identification of CD3 (left column) and IL-17 (right column) in consecutive sections. Abundant IL- 17 staining (arrowhead) is observed in perivascular cells in both acute (A & B) and chronic active (C &D) lesions. (E & F) Whereas CD^{3+} cells do not co-localize with IL- 17 staining within the internal region of a chronic lesion, the fibrillary pattern of IL- 17 immuno-reactivity is representative of possible astrocyte production of IL-17. Minimal staining is observed in tissue from an inactive lesion (G & H). Sparse Th17 cells are observed in NAWM in MS or within control tissue (arrow heads, I-L). Scale bar = 30 μm. Reprinted from Am J Pathol 2008, 172: 146-155 with permission from the American Society for Investigative Pathology.

How might Th17 cells contribute to the pathogenesis of MS? *In vitro* mobilization studies suggest that Th17 cells cross the blood brain barrier (BBB) more efficiently than other T cells. IL-17-secreting CD4 T cells are capable of eliciting damage to the BBB, which would promote greater influx of other inflammatory cells. Once present in the CNS parenchyma, Th17 cells are capable of mediating injury via recruitment of PMNs and monocytes, along with co-production of other cytokines, such as IL-22 and IL-21,. Additionally, Th17 cells are more adept at killing human neurons *in vitro* than un-activated T cells, providing enticing data to suggest that Th17 cells are potential mediators of axonal and neuronal damage in MS lesions. These initial studies support the idea that Th17 cells may be critical mediators of immune destruction of myelin and axons during MS.

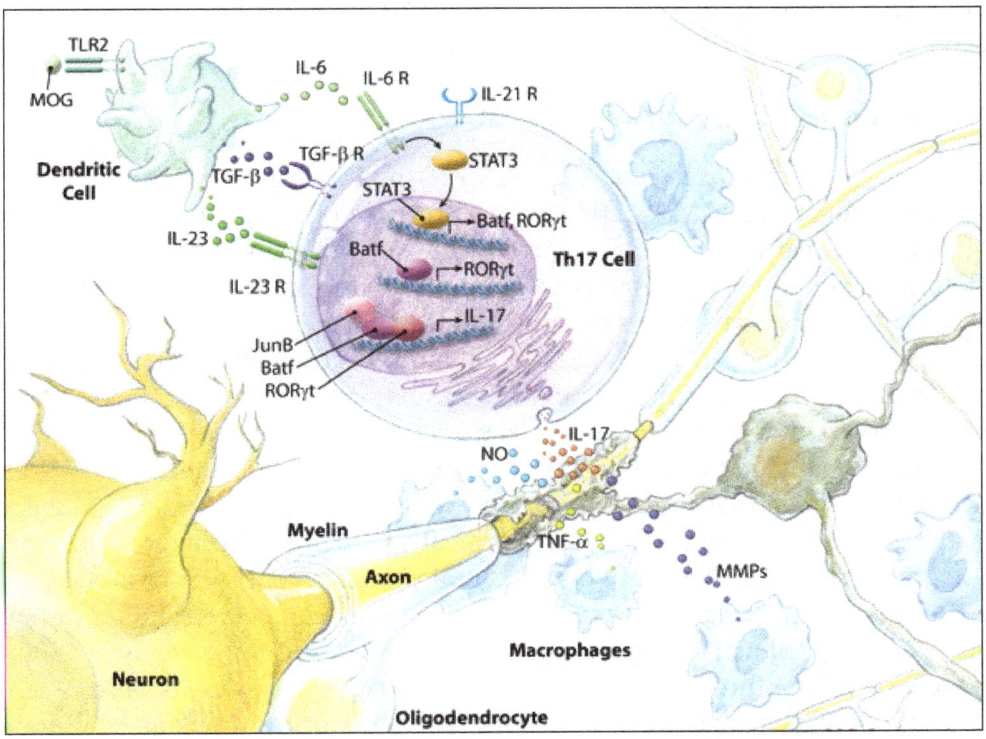

Recent data highlight the potential for targeting Th17 cells in MS. CD4 T cells from healthy individuals skewed to produce IL-17 are more responsive to one current form of treatment for MS, IFNβ, than those from MS patients. Further, IFNβ can inhibit the differentiation of naive CD4 T cells into Th17 cells. Patient responses to IFNβ may be related to the level of IL-17 prior to the onset of therapy. This result re-emphasizes the complexity and heterogeneity of the disease; in particular, the response to current therapies may signify differences in the pathogenesis of MS between patients. The Th17 pathway offers a new therapeutic target, including the molecular pathways that regulate IL-17. However, consistent with the complexity of cytokine networks and regulation, a singular stance that IL-17 is purely pathogenic is likely to be an oversimplification. In particular, IL-17 may participate in limiting tissue destruction during an inflammatory response. Overall, evidence is mounting to suggest that Th17 cells are present during the inflammatory destruction of tissue in MS, but questions remain about the direct contribution of this newly identified lineage of CD4 T cells. At this point, it is reasonable to be inclusive of Th17 cells in the pathogenesis of MS, but to what degree they are involved remains to be shown.

Cellular and molecular factors involved in the immuno-pathogenesis of MS. Rather than representing an all-inclusive summary of the immuno-pathologic features of MS, this diagram highlights recent advances in the understanding of the neural-immune interactions in MS, including factors involved in leukocyte trafficking, axonal injury and antigen presentation. APC = antigen presenting cell; B = B lymphocyte; T = T lymphocyte; Mφ = macrophage, PVMφ = perivascular macrophage; PC = plasma cell; FDC = follicular DC; μglia = microglia; pDC = plasmacytoid DC.

B Cell-mediated CNS Damage in MS

Evidence gathered from examination of CNS tissue implicates the role of antibodies during the pathogenesis of MS. As noted, the presence of plasma cells, Ig and complement is a typical feature of the MS plaque. Naturally, this observation has prompted consideration over whether Ig present within plaques specifically targets myelin antigens. Molecular features of B cells found within MS plaques demonstrate that B cell receptor genes are modified in a specific way that indicates their evolving response toward specific targets. Further specificity of Ig from MS tissue has been described. MBP-reactive and MOG-reactive Ig has been isolated from CNS tissue of patients with MS. Additionally, *in situ* deposition of MOG-specific antibodies has been detected in MS lesions, along with MOG- and MBP-specific Ig complexed with myelin within macrophages. In contrast, detection of Ig appeared non-specifically dispersed throughout cellular constituents of plaques and NAWM, without significant differences compared to control tissues, suggesting that Ig presence may be secondary to white matter injury rather than antigen-specific B cell activation unique to MS. Thus, while controversy persists as to the antigen-specific nature of B cell involvement in MS, the presence of plasma cells and Ig continues to fuel investigation into their involvement during MS plaque development.

As a representation of B cell involvement in CNS parenchymal damage of MS patients, CSF analysis has afforded several clues on the role of Ig in the pathogenesis of MS. The localized intrathecal production of Ig, referred to as oligoclonal bands (OCB), is detected in over 90% of patients with RRMS, and their absence is associated with reduced severity of disease. While OCB are thought to be a product of clonally-expanded B cells within the CSF compartment, their potential targets remain elusive. Plasma cells from the CSF of patients with inflammatory demyelinating disease produce antibodies that are capable of binding myelin and recognizing MBP. However, a more recent study of the CSF of patients with MS failed to detect IgG from CSF clonal B cells that bound to MBP, PLP or MOG. Analysis of OCB has also included investigation of IgM, which is found in a subset of MS patients. CSF-restricted IgM isolated from MS patients have been found to target a variety of lipid antigens, predominantly phosphatidylcholine. These persistent lipid-specific IgM OCBs are associated with more aggressive disease, and might be associated with a poor response to interferon therapy. Taken as a whole, a multitude of potential targets, including those from myelin as well as other CNS antigens, have been proposed and investigated in patients with MS.

Modulation of T cell function may be an equally important function of B cells in the immune dysregulation in MS patients. As noted, anti-CD20 monoclonal antibody targeted depletion of B cells is efficacious for the treatment of RR-MS. This appears independent of antibody effects, since anti-CD20 treatment does not directly target Ig-secreting plasma cells; results in early efficacy in relapse reduction and inflammatory MRI lesions; and does not alter CSF Ig levels. B cells may promote neuro-inflammation in MS via direct and indirect effects on T cells, such as the secretion of the pro-inflammatory cytokines. For example, B cells from MS patients produce more TNF-α and lymphotoxin in the presence of the T cell-derived pro-inflammatory cytokine IFN-γ compared to healthy controls. Following anti-CD20-mediated B cell depletion in RRMS patients T cells produced less IFN-γ and were less proliferative in response to TCR engagement. Conversely, B cells are likely to also have immune-suppressive traits that are important in the immuno-pathogenesis of MS. For example, IL-10 secretion by B cells can serve to limit pro-inflammatory auto-reactive CD4 T cell responses.

Finally, recent evidence suggests a role for B cells in the generation of ectopic follicles in MS. Building on observations of Prineas, Aloisi and colleagues have described the presence of B cell follicles within the meninges in patients with SPMS. These are characterized by features of germinal centers, including the presence of B cells, follicular dendritic cells (FDCs), and CXCL13 (a chemokine involved in genesis of lymphoid organs). Found in approximately 40% of autopsy specimens only from patients with SPMS, the presence of these follicles was purported to be a maker for cortical lesions immediately adjacent to the ectopic follicle within the meninges. This finding has been disputed in subsequent work that failed to detect meningeal follicles in 12 patients with SPMS. Interestingly, there may be an association between the presence of EBV infectious material and secondary lymphoid follicles observed in MS patients. The implication of this finding is that latent infection of B cells with EBV drives the expansion and maturation of B cells along with ectopic follicle generation within the CNS, promoting intrathecal Ig production and the targeted destruction of underlying myelin in this region. While only observed in patients with SPMS, it is possible that the generation of these lymphoid organ-like structures begins during the relapsing-remitting phase and evolves over time. Whether these structures are inhibited by B cell-depleting therapy in MS remains to be investigated.

Mechanisms of Leukocyte Entry during MS

Immune access to the CNS is generally considered restricted. In practice, the traditional view of the CNS as an immune privileged site has been replaced with the more appropriate characterization of the CNS as an immune specialized organ. Thus, one key element to immune-mediated damage within the CNS during MS is the process by which immune cells are able to gain access to this specialized compartment. Within the context of universal processes governing immune cell trafficking, there are features that are relatively specific for leukocyte migration to, and within, the brain and spinal cord in health and disease. The molecular components, location, and timing of migration are all important factors during the immuno-pathogenesis of MS.

The BBB serves to actively restrict cellular and macromolecular movement between the blood and CNS tissue. Adequate function of the BBB depends upon several unique anatomic and cellular features, including tight junctions between endothelial cells, specialized expression of molecular transporters, and placement of immune cells within the CNS relative to the vasculature. Only by engaging in a critically timed sequence of events are auto-reactive lymphocytes able to enter the CNS compartment. Initially, leukocytes engage in rolling, activation and arrest to the endothelium of the BBB. This is greatly facilitated by up-regulation of adhesion molecules by the vasculature, including ICAM1 and VCAM1. While a trigger for vascular change remains unclear in MS, hypothetically changes in the vascular endothelium could result from pro-inflammatory mediators circulating within the vasculature, including TNF and/or LPS. Subsequently, migration of cells through and between endothelial cells takes place. Eventually, concentrated extravasation of immune cells in perivascular cuffs within the CNS parenchyma culminates in a breach of the BBB that is an essential component to the process of inflammatory destruction of white matter in MS.

The complex set of molecules that leukocytes depend upon for entry into CNS tissues involves integrins. Integrins are hetero-dimeric cell surface molecules that mediate adhesion between cells. Out of a panel of leukocyte adhesion receptors, the α4 subunit of VLA-4 was identified as a crucial factor for encephalitogenic T cell binding to CNS endothelium. Blockade of α4β1 engagement with one of its binding partners, VCAM-1, successfully abrogated disease in an animal model of MS Clinical trials of a humanized monoclonal antibody targeting the α4 subunit of VLA- 4, called natalizumab, also demonstrated efficacy in the treatment of MS. Hence, selective inhibition of specific adhesion molecules are effective at reducing leukocyte entry into the CNS. Of note, natalizumab reduces the influx of a wide range of leukocytes, including T cells and dendritic cells (DCs). In addition to VLA-4, other trafficking molecules impart specificity of migration into the CNS. Recently, ALCAM-1 was shown to be localized to the BBB and up-regulated in active MS lesions. In experimental animal systems, blockade of ALCAM-1 delayed disease. In addition, osteopontin also serves as a binding partner for VLA-4 and potentially serves as a separate target for reducing leukocyte migration into the CNS of patients with MS. Hence, a multitude of adhesion molecules participate in effective leukocyte trafficking to and within the CNS and serve as potential targets for therapies in MS.

Chemokines, a broad class of cytokines mediating chemotaxis, also contribute to leukocyte migration to the CNS. Several chemokines and their receptors have been implicated in leukocyte influx to the CNS in MS. For example, CXCL12, constitutively expressed in the CNS, is typically localized to the basolateral aspect of the CNS microvasculature and functions to retain leukocytes within the perivascular space. Redistribution of CXCL12 to the luminal aspect of vessels was observed in autopsy specimens from MS patients, which would allow for the dissemination of lymphocytes into the CNS parenchyma. Other chemokines are thought to participate in the recruitment of lymphocytes into the CNS in MS. Recent work in the EAE system has demonstrated that the initial wave of T cell infiltration into the CNS prior to disease onset is CCR6-dependent. Furthermore, in EAE, the initial wave of inflammatory CD4 T cells express IL-17 and are potentially recruited specifically via CCR6 into the CNS via the choroids plexus that expresses the CCR6 ligand, CCL20. These results have yet to be convincingly replicated in human samples. Several other chemokine and chemokine receptors expressed by various cell types in MS show dysregulation including CCR7, CCL19 and CCL21, CCR5. Thus, there may be a unique chemokine signature at different phases of disease and in different regions of the CNS in order for various leukocytes to localize to the CNS during MS.

Role of Antigen Presenting Cells in MS

Antigen presenting cells (APCs) process and present antigens to T cells, and in the context of MHC, co-stimulatory signals and cytokine secretion, drive adaptive immune responses. Experimental evidence based on animal models has shown that antigen-specific encounters within the CNS between T cells and APCs is crucial to the unfolding of MS. In EAE, without newly generated myelin antigens from the CNS by APCs, inflammatory demyelination does not proceed, even in the presence of myelin-reactive CD4 T cells. Thus, CD4 T cell-mediated disease is thought to be a multi-stage process, involving the initial activation of auto-reactive CD4 T cells, as well as reactivation within the CNS immune compartment. DCs are thought to be the major APC during the secondary phase of cognate interactions with CD4 T cells within the CNS. Perivascular spaces within areas bordering edges of active lesions contain immnostaining for CD209, a marker for a subset of DCs, suggesting that antigen presentation by DCs at the interface of the BBB contributes to the earliest inflammatory processes promoting lesion formation. Visualization of APC interactions with encephalitogenic CD4 T cells using intravital microscopy in a rat model of MS suggests that primed CD4 T cells actively engage with these perivascular APCs en route to entry within the CNS.

In addition to DCs, other APCs likely play an important role in antigen presentation during the pathogenesis of MS. Microglia are hematogenously-derived resident APCs within the CNS. Upon activation, they express greater amounts of MHCII and co-stimulatory molecules , signifying a greater capacity to promote pro-inflammatory T cell responses within the CNS. Activated microglia are localized to active plaques. Experimentally, impeding microglial function attenuates EAE. However, relative to other professional APCs, microglia are not as potent at inducing auto-reactive T cell responses and may even down-regulate CD4 T cell functions. Thus, as resident APCs within the CNS, microglia are capable of performing APC functions, but likely are not the lynchpin for driving autoimmunity of the CNS in MS. Another APC potentially involved in driving myelin-reactive CD4 T cells in MS is the B cell. Recent work suggests that B cells play a prominent role in the pathogenesis of MS, and potentially play important roles in antigen presentation to T cells. In addition to effects on T cells, MHCII-dependent interactions with B cells promotes Ig class switching from IgM to IgG. Thus, antigen-specific interactions between B cells and T cells represent a critical step in the generation of Ig responses in MS. It is important to acknowledge that not all interactions between APCs and T cells promote inflammation. In addition to the effects of regulatory B cells APCs can also engender anti-inflammatory responses. For example, suppressor myeloid cells are generated after EAE induction and are capable of suppressing T cell function. In MS, the process of myeloid suppression is thought to be regulated in part by TREM-2, a trans-membrane signaling protein expressed by microglial cells, macrophages, monocytes and DCs. This mechanism may be dysregulated by secretion of soluble TREM-2, which could act as a decoy receptor and prevent inhibitory function of transmembrane TREM-2.

Axonal and Neuronal Damage in MS

Inflammatory CNS injury in MS has increasingly been associated with axonal damage. Although MS has classically been described as a disease marked by the loss of myelin in greater proportion to the loss of axons, axonal damage was noted in the earliest pathological descriptions of MS lesions. Modern techniques have allowed for precise demonstrations of axonal damage. Antibodies directed at amyloid precursor protein show damaged axons in active areas of MS lesions. Representing a major advance in the field of MS pathology, Trapp *et al* were able to directly view and quantify transected axons using confocal microscopy by counting axonal ovoids at the ends of transected axons. The active areas of MS lesions were found to contain more transected axons than inactive areas in more chronic lesions. Of note, comparisons of biopsy and autopsy samples from patients with relapsing-remitting, secondary progressive, and primary progressive MS suggest that axonal pathology is greatest within the first year of disease onset, particularly in patients with RRMS. These studies propose that axonal pathology occurs in areas of active inflammatory demyelination and early during the course of disease.

In addition, a slower rate of axonal damage is also thought to occur and contribute to the clinical decline observed in MS patients. Trapp *et al* and Kornek *et al* showed that axonal ovoids are more common in inactive demyelinated lesions and in NAWM than in the white matter of control patients. However, remyelinated inactive lesions, or shadow plaques, have the same number of abnormal axons as control tissue. Patients with higher levels of motor disability have fewer surviving corticospinal axons traveling through their spinal cord, demonstrating a direct correlation of axonal damage and disease progression.

The mechanisms involved in axonal damage in MS are under intense investigation. CD8+ T-lymphocytes can cause axonal damage via the release of cytotoxic granules, induction of apoptosis through activation of surface receptors such as Fas, the release of cytokines such as TNF-α, or direct transection of axons. Macrophages/microglia are also found in close proximity to damaged neurons. Release of toxic molecules by these cells such as proteases and reactive nitrogen species can cause oligodendrocyte injury, demyelination, and axonal degeneration, disrupt the blood-brain barrier, and contribute to the loss of axonal conduction. Axonal damage also occurs by activation of other components of the innate immune response such as Toll-like receptors. Toll-like receptor 2 is over-expressed by oligodendrocytes in MS lesions where it inhibits remyelination. Antibody-mediated injury to axonal components, e.g. neurofascin, can result in axonal and neuronal dysfunction. As a consequence of immune injury to myelin, higher energy demands on demyelinated axons and glutamate-mediated excitotoxicity may further impart unsustainable damage. Overall, axonal injury in MS is likely mediated by multiple mechanisms at play in both active and chronic lesions.

Neuronal loss in MS can be severe and occurs throughout the brain. Neuronal loss in the range of 18–35% has been reported in the cortex, hippocampus, thalamus, and spinal cord (reviewed in). Damage to axons and neurons has been evaluated *in vivo* using MRI techniques such as quantitative proton MR spectroscopy. In patients presenting with their first clinical attack of MS, the amount of NAA in the whole brain is already decreased, indicating early neuronal damage. This reduction is still present one year later and was independent of whether the patients had progressed to develop MS. Other MRI techniques, such as measurement of brain volume or diffusion tensor imaging, have provided more variable results in evaluating axonal and neuronal damage especially in short term studies as factors such as demyelination and edema can confound the results. However, examination of neuronal integrity using OCT shows a loss of macular volume in patients with progressive forms of MS, which correlates with poor visual acuity especially in patients with a history of optic neuritis. Similar to axonal injury, the processes resulting in neuronal loss in MS are likely several-fold. Direct immune injury to the gray matter can result in a loss of neurons, as within gray matter lesions a significant increase in apoptotic neurons was observed primarily in large pyramidal cortical neurons. However, in layer II of primary motor cortex from NAWM, paralbumin interneurons were more affected than other neurons that are relatively spared in MS patients. This differential susceptibility of neurons exposed to the same insult is part of a key consideration in how clinical deterioration, particularly with secondary progression, is related to repeated accumulation of axonal or neuronal damage to various neuronal populations. Further, axonal and neuronal survival may be directly tied to the trophic support provided by myelin, which may be particularly relevant during a high metabolic demand state of neurons exposed to inflammatory stressors.

References

- Hall, John E.; Guyton, Arthur C. (2006). Textbook of medical physiology. St. Louis, Mo: Elsevier Saunders. p. 228. ISBN 0-7216-0240-1
- Holmes, Nick (15 November 2001). "Part 1B Pathology: Lecture 11 - The Complement System". Archived from the original on 9 January 2006. Retrieved 2006-05-10
- What-is-Pathology, health: news-medical.net, Retrieved 26 April, 2020
- Doitsh, G; Greene, WC (2016). "Dissecting How CD4 T Cells Are Lost During HIV Infection". Cell Host Microbe. 19 (3): 280–91. doi:10.1016/j.chom.2016.02.012. PMC 4835240. PMID 26962940
- Holmes, Nick (15 November 2001). "Part 1B Pathology: Lecture 11 - The Complement System". Archived from the original on 9 January 2006. Retrieved 2006-05-10
- Lumeng, Carey N.; Saltiel, Alan R. (2011-06-01). "Inflammatory links between obesity and metabolic disease". Journal of Clinical Investigation. 121 (6): 2111–2117. doi:10.1172/jci57132. ISSN 0021-9738. PMC 3104776. PMID 21633179

4

Medical Genetics

Medical genetics is the branch of medicine that is concerned with the management and diagnosis of hereditary disorders. Genetic engineering, gene therapy, genetic testing and counseling, etc. are studied under its domain. This chapter closely examines medical genetics to provide an extensive understanding of the subject.

Medical geneticists specialize in medicine that involves the interaction between genes and health. They are trained to evaluate, diagnose, manage, treat and counsel individuals of all ages with hereditary disorders. This specialist uses modern cytogenetic, molecular, genomic and biochemical genetic testing to assist in specialized diagnostic evaluations, implement needed therapeutic interventions and provide genetic counseling and prevention through prenatal and preimplantation diagnosis. The Medical Geneticist plans and coordinates screening for genetic diseases involving single gene and chromosomal disorders, congenital anomalies, inborn errors of metabolism, multifactorial conditions, and common disorders with hereditary factors.

Primary Speciality Certificates

- Clinical Biochemical Genetics: A clinical biochemical geneticist demonstrates competence in directing and interpreting a wide range of specialized, laboratory biochemical genetic analyses relevant to the diagnosis and management of inherited metabolic disorders. The specialist acts as a consultant regarding laboratory diagnosis on a broad range of inborn errors of metabolism.

- Clinical Genetics and Genomics (MD): A clinical geneticist demonstrates competence in providing comprehensive diagnostic, management, therapeutic, and counseling services for individuals and families at risk for clinical disorders with a genetic basis. The specialist is trained to evaluate individuals of all ages for hereditary conditions.

- Laboratory Genetics and Genomics: A diplomate certified in Laboratory Genetics and Genomics demonstrates competence in directing and interpreting both clinical cytogenetic and molecular genetic analyses relevant to the diagnosis and management of human genetic disease. This specialist acts as a consultant in laboratory diagnoses for a broad range of molecular and chromosomal-based disorders, including both inherited and acquired conditions.

Subspecialties

- Medical Biochemical Genetics: A medical biochemical geneticist demonstrates competence in the diagnosis, medical treatment and management of individuals with inherited metabolic conditions presenting clinically from infancy through adulthood, including via newborn screening. The subspecialist provides direct care and consultative care for individuals of all ages who are diagnosed with inborn errors of metabolism.
- Molecular Genetic Pathology: A molecular genetic pathologist is expert in the principles, theory and technologies of molecular biology and molecular genetics. This expertise is used to make or confirm diagnoses of Mendelian genetic disorders, of human development, infectious diseases and malignancies and to assess the natural history of those disorders. A molecular genetic pathologist provides information about gene structure, function and alteration and applies laboratory techniques for diagnosis, treatment and prognosis for individuals with related disorders.

GENETIC ENGINEERING

Genetic engineering is the artificial manipulation, modification, and recombination of DNA or other nucleic acid molecules in order to modify an organism or population of organisms.

Genetic engineering.

A genetically engineered salmon (top) and a natural salmon of the same age (bottom). The ability to engineer and precisely edit the genomes of animals, while potentially beneficial, has raised ethical questions.

Process and Techniques

Most recombinant DNA technology involves the insertion of foreign genes into the plasmids of common laboratory strains of bacteria. Plasmids are small rings of DNA; they are not part of the bacterium's chromosome (the main repository of the organism's genetic information). Nonetheless, they are capable of directing protein synthesis, and, like chromosomal DNA, they are reproduced and passed on to the bacterium's progeny. Thus, by incorporating foreign DNA (for example, a mammalian gene) into a bacterium, researchers can obtain an almost limitless number of copies of the inserted gene. Furthermore, if the inserted gene is operative (i.e., if it directs protein synthesis), the modified bacterium will produce the protein specified by the foreign DNA.

A subsequent generation of genetic engineering techniques that emerged in the early 21st century centred on gene editing. Gene editing, based on a technology known as CRISPR-Cas9, allows researchers to customize a living organism's genetic sequence by making very specific changes to its DNA. Gene editing has a wide array of applications, being used for the genetic modification of crop plants and livestock and of laboratory model organisms (e.g., mice). The correction of genetic errors associated with disease in animals suggests that gene editing has potential applications in gene therapy for humans.

Applications

Genetic engineering has advanced the understanding of many theoretical and practical aspects of gene function and organization. Through recombinant DNA techniques, bacteria have been created that are capable of synthesizing human insulin, human growth hormone, alpha interferon, a hepatitis B vaccine, and other medically useful substances. Plants may be genetically adjusted to enable them to fix nitrogen, and genetic diseases can possibly be corrected by replacing dysfunctional genes with normally functioning genes. Nevertheless, special concern has been focused on such achievements for fear that they might result in the introduction of unfavourable and possibly dangerous traits into microorganisms that were previously free of them—e.g., resistance to antibiotics, production of toxins, or a tendency to cause disease. Likewise, the application of gene editing in humans has raised ethical concerns, particularly regarding its potential use to alter traits such as intelligence and beauty.

GENE THERAPY

Gene therapy (also called human gene transfer) is a medical field which focuses on the utilization of the therapeutic delivery of nucleic acid into a patient's cells as a drug to treat disease. The first attempt at modifying human DNA was performed in 1980 by Martin Cline, but the first successful nuclear gene transfer in humans, approved by the National Institutes of Health, was performed in May 1989. The first therapeutic use of gene transfer as well as the first direct insertion of human DNA into the nuclear genome was performed by French Anderson in a trial starting in September 1990. It is thought to be able to cure many genetic disorders or treat them over time.

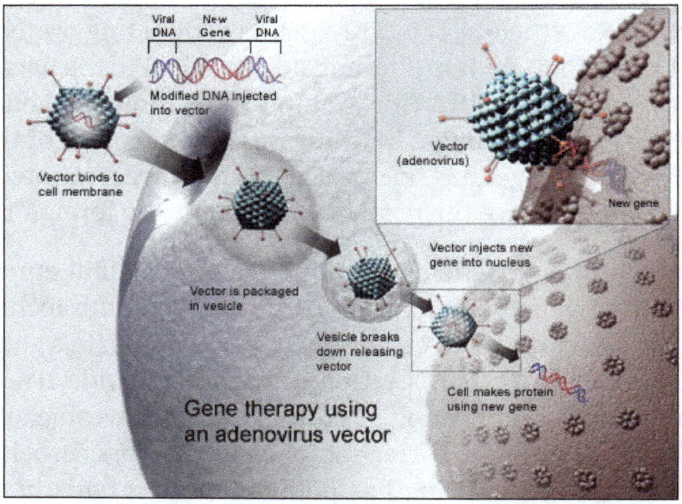

Gene therapy using an adenovirus vector. In some cases, the adenovirus will insert the new gene into a cell. If the treatment is successful, the new gene will make a functional protein to treat a disease.

Between 1989 and December 2018, over 2,900 clinical trials were conducted, with more than half of them in phase I. As of 2017, Spark Therapeutics' Luxturna (RPE65 mutation-induced blindness) and Novartis' Kymriah (Chimeric antigen receptor T cell therapy) are the FDA's first approved gene therapies to enter the market. Since that time, drugs such as Novartis' Zolgensma and Alnylam's Patisiran have also received FDA approval, in addition to other companies' gene therapy drugs. Most of these approaches utilize adeno-associated viruses (AAVs) and lentiviruses for performing gene insertions, *in vivo* and *ex vivo*, respectively. ASO / siRNA approaches such as those conducted by Alnylam and Ionis Pharmaceuticals require non-viral delivery systems, and utilize alternative mechanisms for trafficking to liver cells by way of GalNAc transporters.

The introduction of CRISPR gene editing has opened new doors for its application and utilization in gene therapy. Solutions to medical hurdles, such as the eradication of latent human immunodeficiency virus (HIV) reservoirs and correction of the mutation that causes sickle cell disease, may soon become a tangible reality.

Not all medical procedures that introduce alterations to a patient's genetic makeup can be considered gene therapy. Bone marrow transplantation and organ transplants in general have been found to introduce foreign DNA into patients. Gene therapy is defined by the precision of the procedure and the intention of direct therapeutic effect.

Cell Types

Gene therapy may be classified into two types:

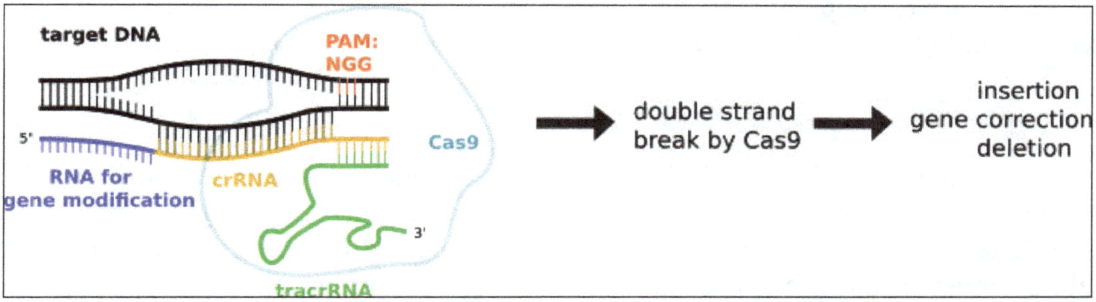

A duplex of crRNA and tracrRNA acts as guide RNA to introduce a specifically located gene modification based on the RNA 5' upstream of the crRNA. Cas9 binds the tracrRNA and needs a DNA binding sequence (5'NGG3'), which is called protospacer adjacent motif (PAM). After binding, Cas9 introduces a DNA double strand break, which is then followed by gene modification via homologous recombination (HDR) or non-homologous end joining (NHEJ).

Somatic

In somatic cell gene therapy (SCGT), the therapeutic genes are transferred into any cell other than a gamete, germ cell, gametocyte, or undifferentiated stem cell. Any such modifications affect the individual patient only, and are not inherited by offspring. Somatic gene therapy represents mainstream basic and clinical research, in which therapeutic DNA (either integrated in the genome or as an external episome or plasmid) is used to treat disease.

Over 600 clinical trials utilizing SCGT are underway in the US. Most focus on severe genetic disorders, including immunodeficiencies, haemophilia, thalassaemia, and cystic fibrosis. Such single gene disorders are good candidates for somatic cell therapy. The complete correction of a genetic disorder or the replacement of multiple genes is not yet possible. Only a few of the trials are in the advanced stages.

Germline

In germline gene therapy (GGT), germ cells (sperm or egg cells) are modified by the introduction of functional genes into their genomes. Modifying a germ cell causes all the organism's cells to contain the modified gene. The change is therefore heritable and passed on to later generations. Australia, Canada, Germany, Israel, Switzerland, and the Netherlands prohibit GGT for application in human beings, for technical and ethical reasons, including insufficient knowledge about possible risks to future generations and higher risks versus SCGT. The US has no federal controls specifically addressing human genetic modification (beyond FDA regulations for therapies in general).

Vectors

The delivery of DNA into cells can be accomplished by multiple methods. The two major classes are recombinant viruses (sometimes called biological nanoparticles or viral vectors) and naked DNA or DNA complexes (non-viral methods).

Viruses

In order to replicate, viruses introduce their genetic material into the host cell, tricking the host's cellular machinery into using it as blueprints for viral proteins. Retroviruses go a stage further by having their genetic material copied into the genome of the host cell. Scientists exploit this by substituting a virus's genetic material with therapeutic DNA. (The term 'DNA' may be an oversimplification, as some viruses contain RNA, and gene therapy could take this form as well.) A number of viruses have been used for human gene therapy, including retroviruses, adenoviruses, herpes simplex, vaccinia, and adeno-associated virus. Like the genetic material (DNA or RNA) in viruses, therapeutic DNA can be designed to simply serve as a temporary blueprint that is degraded naturally or (at least theoretically) to enter the host's genome, becoming a permanent part of the host's DNA in infected cells.

Non-viral

Non-viral methods present certain advantages over viral methods, such as large scale production and low host immunogenicity. However, non-viral methods initially produced lower levels of transfection and gene expression, and thus lower therapeutic efficacy. Newer technologies offer promise of solving these problems, with the advent of increased cell-specific targeting and subcellular trafficking control.

Methods for non-viral gene therapy include the injection of naked DNA, electroporation, the gene gun, sonoporation, magnetofection, the use of oligonucleotides, lipoplexes, dendrimers, and inorganic nanoparticles.

More recent approaches, such as those performed by companies such as Ligandal, offer the possibility of creating cell-specific targeting technologies for a variety of gene therapy modalities, including RNA, DNA and gene editing tools such as CRISPR. Other companies, such as Arbutus Biopharma and Arcturus Therapeutics, offer non-viral, non-cell-targeted approaches that mainly exhibit liver trophism. In more recent years, startups such as Sixfold Bio, GenEdit, and Spotlight Therapeutics have begun to solve the non-viral gene delivery problem. Non-viral techniques offer the possibility of repeat dosing and greater tailorability of genetic payloads, which in the future will be more likely to take over viral-based delivery systems.

Companies such as Editas Medicine, Intellia Therapeutics, CRISPR Therapeutics, Casebia, Cellectis, Precision Biosciences, bluebird bio, and Sangamo have developed non-viral gene editing techniques, however frequently still use viruses for delivering gene insertion material following genomic cleavage by guided nucleases. These companies focus on gene editing, and still face major delivery hurdles.

Moderna Therapeutics and CureVac focus on delivery of mRNA payloads, which are necessarily non-viral delivery problems.

Alnylam, Dicerna Pharmaceuticals, and Ionis Pharmaceuticals focus on delivery of siRNA (antisense oligonucleotides) for gene suppression, which also necessitate non-viral delivery systems.

In academic contexts, a number of laboratories are working on delivery of PEGylated particles, which form serum protein coronas and chiefly exhibit LDL receptor mediated uptake in cells *in vivo*.

Gene Therapy for Osteoarthritis

Gene transfer strategies for the potential medical management of osteoarthritis are under preliminary research to define pathological mechanisms and possible treatments for this chronic disease. Unlike pharmacological treatments which are administered systemically, gene therapy aims to establish sustained, synthesis of gene products and tissue rehabilitation within the joint.

Theory

Passing from parents to children, genes are the building blocks of inheritance. They contain instructions for making proteins. If genes do not produce the right proteins in a correct way, a child can have a genetic disorder. Gene therapy is a molecular method aiming to replace defective or absent genes, or to counteract the ones undergoing overexpression. For this purpose, three techniques may be utilized: gene isolation, manipulations, and transferring to target cells. The most common form of gene therapy involves inserting a normal gene to replace an abnormal gene. Other approaches including repairing an abnormal gene and altering the degree to which a gene is turned on or off. Two basic methodologies are utilized to transfer vectors into target tissues; Ex vivo gene transfer and In-vivo gene transfer. One type of gene therapy in which the gene transfer takes place outside the patient's body is called ex vivo gene therapy. This method of gene therapy is more complicated but safer since it is possible to culture, test, and control the modified cells.

Significance and Causes of Osteoarthritis

Osteoarthritis (OA) is a degenerative joint disease which is the western world's leading cause of pain and disability. It is characterized by the progressive loss of normal structure and function of articular cartilage, the smooth tissue covering the end of the moving bones. This chronic disease not only affects the articular cartilage but the subchondral bone, the synovium and periarticular tissues are other candidates. People with OA can experience severe pain and limited motion. OA is mostly the result of natural aging of the joint due to biochemical changes in the cartilage extracellular matrix.

Osteoarthritis is caused by mechanical factors such as joint trauma and mechanical overloading of joints or joint-instability. Since the degeneration of cartilage is an irreversible phenomenon, it is incurable, costly and responds poorly to treatment. Due to the prevalence of this disease, the repair and regeneration of articular cartilage has become a dominant area of research. The growing number of the people suffering from osteoarthritis and the effectiveness of the current treatments attract a great deal of attention to genetic-based therapeutic methods to treat the progression of this chronic disease.

Vectors for Osteoarthritis Gene Delivery

Various vectors have been developed to carry the therapeutic genes to cells. There are two broad categories of gene delivery vectors: Viral vectors, involving viruses and non-viral agents, such as polymers and liposomes.

Viral Vectors

Viral vectors proved to be more successful in transfecting cells as their life cycles require them to transfer their own genes to the host cells with high efficiency. A virus infects the human by inserting its gene directly into his cells. This can be deadly, but the brilliant idea is to take advantage of this natural ability. The idea is to remove all the dangerous genes in the virus and inject the healthy human genes. So, viruses are inserting positive elements to the host cells while attacking them and they will be helpful rather than harmful.

While viral vectors are 40% more efficient in transferring genes, they are not fully appreciated for in vivo gene delivery because of their further adverse effects. Primarily, viral vectors induce an inflammatory response, which can cause minor side effects such as mild edema or serious ones like multisystem organ failure. It is also difficult to administer gene therapy repeatedly due to the immune system's enhanced response to viruses. Furthermore, viruses may spread out to other organs after intraarticular injection and this will be an important disadvantage. However, majority of problems associated with gene delivery using viral vectors solved by ex vivo gene delivery method. In Osteoarthritis gene therapy, ex vivo method makes it possible to transfect not only the cells of the synovial lining of joints but also articular chondrocytes and chondroprogenitor cells in cartilage.

Non-viral Vectors

Non-viral methods involve complexing therapeutic DNA to various macromolecules including cationic lipids and liposomes, polymers, polyamines and polyethylenimine, and nanoparticles. FuGene 6 and modified cationic liposomes are two non-viral gene delivery methods that have so far been utilized for gene delivery to cartilage. FuGene 6 is a non-liposomal lipid formulation, which has proved to be successful in transfecting a variety of cell lines. Liposomes have shown to be an appropriate candidate for gene delivery, where cationic liposomes are made to facilitate the interaction with the cell membranes and nucleic acids. Unlike viral vectors, non-viral ones avoid the risk of acquiring replication competence. They have the capacity to deliver a large amount of therapeutic genes repeatedly, and it is convenient to produce them on a large scale. The most important of all, they do not elicit immune responses in the host organism. In spite of having advantages, non-viral vectors have not yet replaced viral vectors due to relatively low efficiency and short-term transgene expression.

Novel non-viral vectors for osteoarthritis gene delivery including polymeric vectors are still under investigations.

Target Cells in Osteoarthritis Gene Therapy

Target cells in the OA therapy are autologous chondrocytes, Chondroprogenitor cells, Cells within the synovial cavity, and cells of adjacent tissues such as muscle, tendons, ligaments, and meniscus. Development of cartilage function and structure may be achieved by:

- Inhibiting inflammatory and catabolic pathways.
- Stimulating anabolic pathways to rebuild the matrix.
- Impeding cell senescence.
- Avoiding the pathological formation of osteophytes.
- Prevention of apoptosis, and influencing several of these processes.

Approaches influencing several of these processes simultaneously have also shown to be successful, like transferring the combination of inhibitors of catabolism pathways and activators of anabolic events (IGF-I/IL-1RA), as well as that of activators of anabolic and proliferative processes (FGF- 2/SOX9 or FGF-2/IGF-I).

Gene Defects Leading to Osteoarthritis

Osteoarthritis has a great degree of heritability. Forms of osteoarthritis caused by single gene mutation have better chance of treatment by gene therapy.

Epidemiological studies have shown that a genetic component may be an important risk factor in OA. Insulin-like growth factor I genes (IGF-1), Transforming growth factorβ, cartilage oligomeric matrix protein, bone morphogenetic protein, and other anabolic gene candidates are among the candidate genes for OA. Genetic changes in OA can lead to defects of a structural protein such as collagen, or changes in the metabolism of bone and cartilage. OA is rarely considered as a simple disorder following Mendelian inheritance being predominantly a multifactorial disease.

However, in the field of OA gene therapy, researches has more focused on gene transfer as a delivery system for therapeutic gene products, rather counteracting genetic abnormalities or polymorphisms. Genes, which contribute to protect and restore the matrix of articular cartilage, are attracting the most attention. These Genes are listed in Table. Among all candidates listed below, proteins that block the actions of interleukin-1 (IL-1) or that promote the synthesis of cartilage matrix molecules have received the most experimental scrutiny.

Table: Candidates for OA gene therapy.

Category	Gene Candidate
Cytokine/cytokine antagonist	IL-1Ra, sIL-1R, sTNFR, IL-4
Cartilage growth factor	IGF-1, FGF, BMPs, TGF, CGDF
Matrix breakdown inhibitor	TIMPs, PAIs, serpins
Signaling molecule/transcription factor	Smad, Sox-9, IkB
Apoptosis Inhibitor	Bcl-2
Extra cellular matrix molecule	Type II collagen, COMP
Free radical antagonist	Super Oxide Dismutase

Interleukin-1 as a Target in Osteoarthritis

Researches suggest that among all potential mediators, a protein called Interleukin-1 is by far the most potent cause of the pain, joint inflammation and loss of cartilage associated with osteoarthritis. A therapeutic gene used to treat the arthritic joins produces a second protein, which naturally counteracts the effect of interleukin-1. The Interleukin 1 receptor antagonist (IL-1Ra), the natural agonist of IL-1, is a protein that binds non-productively to the cell surface of interleukin-1 receptor, therefore blocks the activities of IL-1 by preventing it from sending a signal to IL-1 receptor. There are three main researches that prove the benefits of local IL-1Ra gene therapy in animal models of osteoarthritis. Series of experiments on canines, rabbits, and horses demonstrate that local IL-1Ra gene therapy is safe and effective in animal models of OA, according to the fact that recombinant human IL-1Ra strongly protected the articular cartilage from degenerative changes.

Strategies for Osteoarthritis Gene Therapy

In the context of OA, the most attractive intra- articular sites for gene transfer are the synovium and the articular cartilage. Most experimental progress has been made with gene transfer to a convenient intra-articular tissue, such as the synovium, a tissue amenable to genetic modification by a variety of vectors, using both in vivo and ex vivo protocols.

Gene Transfer to Synovium

The major purpose of gene delivery is to alter the lining of the joint in a way that enables them to serve as an endogenous source of therapeutic molecules therapeutic molecules can diffuse and influence the metabolism of adjacent tissues such as cartilage. Genes may be delivered to synovium in animal models of RA and OA by direct, in vivo injection of vector or by indirect, ex vivo methods involving autologous synovial cells, skin fibroblasts, or other cell types such as mesenchymal stem cells.

The direct in vivo approach is intra-articular insertion of a vector to affect synovicytes. Vectors play crucial role in success of this method. The effect of Different vectors for in vivo gene delivery to synovium.

Table: Performance of different vectors for in vivo gene delivery to synovium.

Vector	Comment
Non-viral Vectors	Short-term and less efficient transfection; many inflammatory
Retrovirus	No transduction of normal synovium; modest transduction of inflamed synovium
Lentivirus	Extremely high transduction and transgene expression; no obvious side effects
Adenovirus	High transduction efficiency; dose-dependent inflammatory response
Adeno-associated virus	Moderate levels of transduction of normal and inflamed synovium
Herpes simplex virus	Highly efficient transduction; cytotoxic

The indirect ex vivo approach involves harvest of synovium, isolation and culture of synoviocytes, in vitro transduction, and injection of engineered synovicytes into the joint.

Gene Transfer to Cartilage

Contrary to the synoviocytes which are dividing cells and can be efficacy transduced in vivo using either liposomes or viral vectors, in vivo delivery of genes to chondrocytes is hindered by the dense extra cellular matrix that surrounds these cells. Chondrocytes are non- dividing cells, embedded in a network of collagens and proteoglycans; however researches suggest that genes can be transferred to chondrocytes within normal cartilage by intraarticular injection of liposomes containing sendai virus (HVJ- liposomes) and adeno-associated virus.

Most efficient methods of gene transfer to cartilage have involved ex vivo strategies using chondrocytes or chondroprogenitor cells. Chondrocytes are genetically enhanced by transferring complementary DNA encoding IL-1RA, IGF-1, or matrix break down inhibitors mentioned in Table. As discussed before, the transplanted cells could serve as an intra- articular source of therapeutic molecules.

Safety

One important issue related to human gene therapy is safety, particularly for the gene therapy of non-fatal diseases such as OA. The main concern is the high immunogenicity of certain viral vectors. Retroviral vectors integrate into the chromosomes of the cells they infect. There will be always a chance of integrating into a tumor suppressor gene or an oncogene, leading to virulent transformation of the cell. In general, gene transfer to humans is considered as a safe therapeutic method, despite recent events that have provided examples of random adverse events.

Gene Therapy of the Human Retina

Retinal gene therapy holds a promise in treating different forms of non-inherited and inherited blindness.

In 2008, three independent research groups reported that patients with the rare genetic retinal disease Leber's congenital amaurosis had been successfully treated using gene therapy with adeno-associated virus (AAV). In all three studies, an AAV vector was used to deliver a functional copy of the RPE65 gene, which restored vision in children suffering from LCA. These results were widely seen as a success in the gene therapy field, and have generated excitement and momentum for AAV-mediated applications in retinal disease.

In retinal gene therapy, the most widely used vectors for ocular gene delivery are based on adeno-associated virus. The great advantage in using adeno-associated virus for the gene therapy is that it poses minimal immune responses and mediates long-term transgene expression in a variety of retinal cell types. For example, tight junctions that form the blood-retina barrier, separate subretinal space from the blood supply, providing protection from microbes and decreasing most immune-mediated damages.

Mechanism

Physiological Components in Retinal Gene Therapy

The vertebrate neural retina composed of several layers and distinct cell types. A number of these cell types are implicated in retinal diseases, including retinal ganglion cells, which degenerate in glaucoma, the rod and cone photoreceptors, which are responsive to light and degenerate in retinitis pigmentosa, macular degeneration, and other retinal diseases, and the retinal pigment epithelium (RPE), which supports the photoreceptors and is also implicated in retinitis pigmentosa and macular degeneration.

In retinal gene therapy, AAV is capable of "transducing" these various cell types by entering the cells and expressing the therapeutic DNA sequence. Since the cells of the retina are non-dividing, AAV continues to persist and provide expression of the therapeutic DNA sequence over a long time period that can last several years.

AAV Tropism and Routes of Administration

AAV is capable of transducing multiple cell types within the retina. AAV serotype 2, the most well-studied type of AAV, is commonly administered in one of two routes: intravitreal or subretinal. Using the intravitreal route, AAV is injected in the vitreous humor of the eye. Using the subretinal route, AAV is injected underneath the retina, taking advantage of the potential space between the photoreceptors and RPE layer, in a short surgical procedure. Although this is more invasive than the intravitreal route, the fluid is absorbed by the RPE and the retina flattens in less than 14 hours without complications. Intravitreal AAV targets retinal ganglion cells and a few Muller glial cells. Subretinal AAV efficiently targets photoreceptors and RPE cells.

The reason that different routes of administration lead to different cell types being transfected (e.g., different tropism) is that the inner limiting membrane (ILM) and the various retinal layers act as physical barriers for the delivery of drugs and vectors to the deeper retinal layers. Thus overall, subretinal AAV is 5-10 times more efficient than delivery using the intravitreal route.

Tropism Modification and Novel AAv Vectors

One important factor in gene delivery is developing altered cell tropisms to narrow or broaden rAAV-mediated gene delivery and to increase its efficiency in tissues. Specific properties like capsid conformation, cell targeting strategies can determine which cell types are affected and also the efficiency of the gene transfer process. Different kinds of modification can be undertaken. For example, modification by chemical, immunological or genetic changes that enables the AAV2 capsid to interact with specific cell surface molecules.

Initial studies with AAV in the retina have utilized AAV serotype 2. Researchers are now beginning to develop new variants of AAV, based on naturally-occurring AAV serotypes and engineered AAV variants.

Several naturally-occurring serotypes of AAV have been isolated that can transduce retinal cells. Following intravitreal injection, only AAV serotypes 2 and 8 were capable of transducing retinal ganglion cells. Occasional Muller cells were transduced by AAV serotypes 2, 8, and 9. Following subretinal injection, serotypes 2, 5, 7, and 8 efficiently transduced photoreceptors, and serotypes 1, 2, 5, 7, 8, and 9 efficiently transduce RPE cells.

One example of an engineered variant has recently been described that efficiently transduces Muller glia following intravitreal injection, and has been used to rescue an animal model of aggressive, autosomal-dominant retinitis pigmentosa.

AAV and Immune Privilege in the Retina

Importantly, the retina is immune-privileged, and thus does not experience a significant inflammation or immune-response when AAV is injected. Immune response to gene therapy vectors is what has caused previous attempts at gene therapy to fail, and is considered a key advantage of gene therapy in the eye. Re-administration has been successful in large animals, indicating that no long-lasting immune response is mounted.

Recent data indicates that the subretinal route may be subject to a greater degree of immune privilege compared to the intravitreal route.

Promoter Sequence

Expression in various retinal cell types can be determined by the promoter sequence. In order to restrict expression to a specific cell type, a tissue-specific or cell-type specific promoter can be used.

For example, in rats the murine rhodopsin gene drive the expression in AAV2, GFP reporter product was found only in rat photoreceptors, not in any other retinal cell type or in the adjacent RPE after subretinal injection. On the other hand, if ubiquitously expressed immediate-early cytomegalovirus (CMV) enhancer-promoter is expressed in a wide variety of transfected cell types. Other ubiquitous promoters such as the CBA promoter, a fusion of the chicken-actin promoter and CMV immediate-early enhancer, allows stable GFP reporter expression in both RPE and photoreceptor cells after subretinal injections.

Modulation of Expression

Sometimes modulation of transgene expression may be necessary since strong constitutive expression of a therapeutic gene in retinal tissues could be deleterious for long-term retinal function. Different methods have been utilized for the expression modulation. One way is using exogenously regulatable promoter system in AAV vectors. For example, the tetracycline-inducible expression system uses a silencer/transactivator AAV2 vector and a separate inducible doxycycline-responsive coinjection. When induction occurs by oral doxycycline, this system shows tight regulation of gene expression in both photoreceptor and RPE cells.

Examples and Animal Models

Targeting RPE

One study that was done by Royal College of Surgeons (RCS) in rat model shows that a recessive mutation in a receptor tyrposine kinase gene, mertk results in a premature stop codon and impaired phagocytosis function by RPE cells. This mutation causes the accumulation of outer segment debris in the subretinal space, which causes photoreceptor cell death. The model organism with this disease received a subretinal injection of AAV serotype 2 carrying a mouse Mertk cDNA under the control of either the CMV or RPE65 promoters. This treatment was found to prolong photoreceptor cell survival for several months and also the number of photoreceptor was 2.5 fold higher in AAV-Mertk- treated eyes compared with controls 9 weeks after injection, also they found decreased amount of debris in the subretinal space.

The protein RPE65 is used in the retinoid cycle where the all-trans-retinol within the rod outer segment is isomerized to its 11-cis form and oxidized to 11-cis retinal before it goes back to the photoreceptor and joins with opsin molecule to form functional rhodopsin. In animal knockout model (RPE65-/-), gene transfer experiment shows that early intraocular delivery of human RPE65 vector on embryonic day 14 shows efficient transduction of retinal pigment epithelium in the RPE65-/- knockout mice and rescues visual functions. This shows successful gene therapy can be attributed to early intraocular deliver to the diseased animal.

Targeting of Photoreceptors

Juvenile retinoschisis is a disease that affects the nerve tissue in the eye. This disease is an X-linked recessive degenerative disease of the central macula region, and it is caused by mutation in the RSI gene encoding the protein retinoschisin. Retinoschisin is produced in the photoreceptor and bipolar cells and it is critical in maintaining the synaptic integrity of the retina.

Specifically the AAV 5 vector containing the wild-type human RSI cDNA driven by a mouse opsin promoter showed long-term retinal functional and structural recovery. Also the retinal structural reliability improved greatly after the treatment, characterized by an increase in the outer nuclear layer thickness.

Retinitis Pigmentosa

Retinitis pigmentosa is an inherited disease which leads to progressive night blindness and loss of peripheral vision as a result of photoreceptor cell death. Most people who suffer from RP are born with rod cells that are either dead or dysfunctional, so they are effectively blind at nighttime, since these are the cells responsible for vision in low levels of light. What follows often is the death of cone cells, responsible for color vision and acuity, at light levels present during the day. Loss of cones leads to full blindness as early as five years old, but may not onset until many years later. There have been multiple hypotheses about how the lack of rod cells can lead to the death of cone cells. Pinpointing a mechanism for RP is difficult because there are more than 39 genetic loci and genes correlated with this disease. In an effort to find the cause of RP, there have been different gene therapy techniques applied to address each of the hypotheses.

Different types of inheritance can attribute to this disease; autosomal recessive, autosomal dominant, X-linked type, etc. The main function of rhodopsin is initiating the phototransduction cascade. The opsin proteins are made in the photoreceptor inner segments, then transported to the outer segment, and eventually phagocytized by the RPE cells. When mutations occur in the rhodopsin the directional protein movement is affected because the mutations can affect protein folding, stability, and intracellular trafficking. One approach is introducing AAV-delivered ribozymes designed to target and destroy a mutant mRNA.

The way this system operates was shown in animal model that have a mutant rhodopsin gene. The injected AAV-ribozymes were optimized in vitro and used to cleave the mutant mRNA transcript of P23H (where most mutation occur) in vivo.

Another mutation in the rhodopsin structural protein, specifically peripherin 2 which is a membrane glycoprotein involved in the formation of photoreceptor outersegment disk, can lead to recessive RP and macular degeneration in human. In a mouse experiment, AAV2 carrying a wild-type peripherin 2 gene driven by a rhodopsin promoter was delivered to the mice by subretinal injection. The result showed improvement in photoreceptor structure and function which was detected by ERG (electroretinogram). The result showed improvement of photoreceptor structure and function which was detected by ERG. Also peripherin 2 was detected at the outer segment layer of the retina 2 weeks after injection and therapeutic effects were noted as soon as 3 weeks after injection. A well-defined outer segment containing both peripherin2 and rhodopsin was present 9-month after injection.

Since apoptosis can be the cause of photoreceptor death in most of the retinal dystrophies. It has been known that survival factors and antiapoptoic reagents can be an alternative treatment if the mutation is unknown for gene replacement therapy. Some scientists have experimented with treating this issue by injecting substitute trophic factors into the eye. One group of researchers injected the rod derived cone viability factor (RdCVF) protein (encoded for by the Nxnl1 (Txnl6) gene) into the eye of the most commonly occurring dominant RP mutation rat models. This treatment demonstrated success in promoting the survival of cone activity, but the treatment served even more significantly to prevent progression of the disease by increasing the actual function of the cones. Experiments were also carried out to study whether supplying AAV2 vectors with cDNA for glial cell line-derived neurotrophic factor (GDNF) can have an anti-apoptosis effect on the rod cells. In looking at an animal model, the opsin transgene contains a truncated protein lacking the last 15 amino acids of the C terminus, which causes alteration in rhodopsin transport to the outer segment and leads to retinal degeneration. When the AAV2-CBA-GDNF vector is administered to the subretinal space, photoreceptor stabilized and rod photoreceptors increased and this was seen in the improved function of the ERG analysis. Successful experiments in animals have also been carried out using ciliary neurotrophic factor (CNTF), and CNTF is currently being used as a treatment in human clinical trials.

AAV based Treatment for Retinal Neovascular Diseases

Ocular neovascularization (NV) is the abnormal formation of new capillaries from already existing blood vessels in the eye, and this is a characteristics for ocular diseases such as diabetic retinopathy (DR), retinopathy of prematurity (ROP) and (wet form) age-related macular degeneration (AMD). One of the main players in these diseases is VEGF (Vascular endothelial growth factor) which is known to induce vessel leakage and which is also known to be angiogenic. In normal tissues VEGF stimulates endothelial cell proliferation in a dose dependent manner, but such activity is lost with other angiogenic factors.

Many angiostatic factors have been shown to counteract the effect of increasing local VEGF. The naturally occurring form of soluble Flt-1 has been shown to reverse neovascularization in rats, mice, and monkeys.

Pigment epithelium-derived factor (PEDF) also acts as an inhibitor of angiogenesis. The secretion of PEDF is noticeably decreased under hypoxic conditions allowing the endothelial mitogenic activity of VEGF to dominate, suggesting that the loss of PEDF plays a central role in the development of ischemia-driven NV. One clinical finding shows that the levels of PEDF in aqueous humor of human are decreased with increasing age, indicating that the reduction may lead to the development of AMD. In animal model, an AAV with human PEDF cDNA under the control of the CMV promoter prevented choroidal and retinal NV.

The finding suggests that the AAV-mediated expression of angiostatic factors can be implemented to treat NV. This approach could be useful as an alternative to frequent injections of recombinant protein into the eye. In addition, PEDF and sFlt-1 may be able to diffuse through sclera tissue, allowing for the potential to be relatively independent of the intraocular site of administration.

IMMUNOGENETICS

Immunogenetics or immungenetics is the branch of medical genetics that explores the relationship between the immune system and genetics.

Autoimmune diseases, such as type 1 diabetes, are complex genetic traits which result from defects in the immune system. Identification of genes defining the immune defects may identify new target genes for therapeutic approaches. Alternatively, genetic variations can also help to define the immunological pathway leading to disease.

Its origin is usually attributed to Edward Jenner, who discovered in 1796 that cowpox, or vaccinia, induced protection against human smallpox. It took two centuries before the World Health Organization announced in 1979 that smallpox had been eradicated.

The term immunogenetics is based on the two words immunology and genetics, and is defined as "a sub-category of genetics focussing on the genetic basis of the immune reaction" according to MeSH. Genetics is the science researching the transfer of characteristics from one generation to the next. The genes of an organism (strands of DNA) and the transfer of genes from the parent to the child generation of an organism in the scope of possible variations are the basis of genetics.

Immunology deals with the biological and biochemical basis for the body's defense against germs (such as bacteria, viruses, and fungi), as well as against foreign agents such as biological toxins and environmental pollutants, and failures and malfunctions of these defence mechanisms. Apart from these external effects on the organism, there are also defence reactions regarding the body's own cells, e.g. in the scope of the bodily reactions on cancer and the lacking reaction of a body on healthy cells in the scope of an immune-mediated disease. Hence, immunology is a sub-category of biology.

The term immunogenetics comprises all processes of an organism, which are, on the one hand, controlled and influenced by the genes of the organism, and are, on the other hand, significant with regard to the immunological defence reactions of the organism.

The medical history of immunology and the immune system dates back to the 19th century. However, the prevention and early defence of diseases was an essential task for all shamans, medicine men and early "doctors" during the entire human evolution. The first Nobel Prize in the field of immunogenetics was awarded to Baruj Benacerraf, Jean Dausset and George Davis Snell in 1980 for discovering genetically determined cellular surface structures, which control immunological reactions.

Immunogenetics and Inherited Risk of Multifactorial Diseases

HLA Loci

Genetic studies have provided evidence for association of loci on HLA and:

- Autoimmune and inflammatory diseases: Acute anterior uveitis, alopecia areata, asthma, atopic dermatitis, eczema, rheumatoid arthritis, Behçet's disease, celiac disease, collagenous colitis, granulomatosis with polyangiitis (Wegener granulomatosis), generalized vitiligo, IgA nephropathy, primary biliary cirrhosis, psoriasis, ankylosing spondylitis, systemic lupus erythematosus, vasculitis, type 1 diabetes, Crohn's disease, ulcerative colitis, dermatomyositis, and Graves' disease.
- Infections: Human immunodeficiency virus (HIV) set-point viral load (spVL), HIV-1 control, acquired immunodeficiency syndrome (AIDS) progression, chronic hepatitis B infection and viral clearance, hepatitis B, hepatitis B and C virus-related hepatocellular carcinoma, hepatitis B-related liver cirrhosis, chronic hepatitis C infection, human papillomavirus (HPV) seropositivity, dengue shock syndrome, leprosy, M. tuberculosis infection, malaria, resistance to enteric fever, and visceral leishmaniasis.

- Gastrointestinal diseases: Barrett's esophagus.
- Neurological disorders: Parkinson's disease, narcolepsy, juvenile myoclonic epilepsy, spinocerebellar ataxia, myasthenia gravis, and multiple sclerosis.
- Psychiatric disorders: Schizophrenia and autism.
- Joint diseases: Knee osteoarthritis.
- Cancers of the nasopharynx, cervix, colorectum, lung, blood cells, and bone marrow (lymphoid cancers).
- Adverse drug reactions: Stevens-Johnson syndrome/toxic epidermal necrolysis (carbamazepine), agranulocytosis (clozapine), pancreatitis (thiopurine), and liver injury (terbinafine, fenofibrate, ticlopidine, and pazopanib).
- Response to vaccines: Hepatitis B.
- Male infertility due to nonobstructive azoospermia.

Non-HLA Loci

Generally, loci on non-HLA genes are involved in genetic predisposition to a variety of autoimmune and inflammatory disorders, among which the most associated have been identified in three genes: cytotoxic T-lymphocyte-associated antigen 4 (CTLA4), protein tyrosine phosphatase (PTPN22), and tumor necrosis factor-α (TNF). In particular, it is noticeable that patients who receive hematopoietic stem cell transplantation (SCT) from matched sibling donor might develop acute graft-versus-host disease (GVHD). This reflects that non-HLA components have a part in making the immunogenetic profile that needs special attention in patients scheduled to undergo stem cell transplantation.

Immunogenetics and the Spectrum of Immune Disorders

By engaging pattern recognition receptors (PRRs)—including Toll-like receptors (TLRs) and nucleotide-binding oligomerization domain-like receptors (NLRs)—and signal-transducing molecules, e.g., interleukin-1 receptor-associated kinase 4 (IRAK4), the innate immune system forms the center of recognition of molecular patterns and therefore the first line of defense against foreign antigens. Abnormally low activity of this system causes underdetection of foreign agents that makes individuals susceptible to being infected, while the unwanted action of this system is being reactive to self-components, which is seen in autoimmune situations. In this manner, if anything hinders the proper functioning of the immune system, for example, genetic factors, then it is most likely that the body is prone to autoimmune and infectious diseases.

Immunogenetics aims to cover HLA and non-HLA effects for all the aforementioned categories with emphasis on autoimmune disorders and infections, two ends of the spectrum of immune disorders.

Autoimmune Diseases

Psoriasis

Psoriasis was initially known as a disease of abnormal keratinocyte proliferation presented with chronic plaque in the majority of instances that can predispose patients to cardiovascular, psychiatric, and joint complications. It is already considered an immune-mediated skin disease where both innate and adaptive immunities play a role in initiating psoriatic lesions. Of note, Th1 pathway is overstimulated; there are high levels of Th1 cytokines and chemokines including IL-2, IL-12, and IFN-γ in psoriatic plaques. T cells, natural killer cells, natural killer T cells, and, to a lesser extent, neutrophils contribute to cutaneous inflammation in psoriasis as well. More interestingly, dendritic cells through antigen presentation to T cells can lead to plaque formation.

Consistent with its different clinical facets, there is a catalog consisting more than 60 loci identified as risk genes for psoriasis. In particular, the risk allele at MHC class I gene is present in about half of patients with psoriasis. Generally, HLA and non-HLA genetic loci associated with psoriasis are important mediators of antigen presentation, Th17/IL-23 axis, T-cell function, antiviral immunity, macrophage activation, and nuclear factor-κB (NF-κB)-dependent signaling. A simple mechanism for these genetic risk alleles in pathogenesis of psoriasis is likely to be mediated through reducing the threshold for activation of the innate immunity.

Rheumatoid Arthritis (RA)

RA is a chronic inflammatory disease of synovial joints characterized by synovial hyperplasia, production of autoantibodies such as rheumatoid factor (RF) and anti-citrullinated protein antibody (ACPA), bone deformity, and systemic manifestations. Collectively, RA is associated with unfavorable long-term prognosis. Different cells such as macrophages, monocytes, fibroblasts, and T cells act to make an orchestra of cytokines, e.g., IL-1, IL-6, and TNF-α, which are central to the abnormal signaling pathways underlying this inflammatory arthritis.

RA has a long-recognized association with alleles at MHC class II gene that contain a common amino acid sequence in the HLA-DRB1 region, e.g., HLA-DRB1*0404 and DRB1*0401. In addition, there are non-HLA genetic loci associated with RA in ACPA-positive patients. As reviewed in, RA-associated genes are known to play a role in the nuclear factor-κB (NF-κB)-dependent signaling, TCR signaling, and JAK-STAT signaling.

Autoimmune Thyroid Diseases (AITDs)

As for other autoimmune disorders, AITDs such as Graves's disease and Hashimoto's thyroiditis are of T-cell-mediated autoimmune disorders mainly characterized by production of autoantibodies against and T-cell infiltration in the thyroid gland. Thereby, the immune system cannot correctly maintain a constant battle causing the gland to malfunction, which can be manifested as hyper- or hypothyroidism. Generally, AITDs are of special importance in practice because of their comorbidities with other autoimmune diseases.

Among risk alleles for AITDs are genetic variants associated with thyroid function. However, a variety of AITDs-related genetic loci occur within immune-modulating and HLA genes, which are known to contribute to peripheral tolerance, T-lymphocyte activation, and antigen presentation. In this manner, a mechanism of immunogenetic susceptibility to AITDs is maintained through interference with central and peripheral tolerance, APCs, and subsequent activation of T cells.

Primary Biliary Cirrhosis (PBC)

It is a disease of small intrahepatic bile ducts regarded as an autoimmune liver disease with the presence of the antimitochondrial antibody (AMA) in all except a minority of patients (up to 10%). Supporting this is that the infiltration of autoreactive CD4+ T cells and CD8+ T cell specific to AMA has increased manifold in the liver of patients with PBS. T cells along with other immune cells such as B cells, macrophages, eosinophils, and natural killer cells take part in the composition of the portal inflammation. Eventually, such a chronic inflammation would progress to the loss of biliary epithelial cells.

Genetic studies have detected HLA variants conferring susceptibility to PBC. However, there were HLA variants that seemed protective against PBC. Non-HLA loci associated with PBC mainly involve genes associated with T cells. In particular, they contribute to IL-12-JAK-STAT4, CD80/CD86, and IL7R-α/CD127 signaling pathways, which are known to play a role in Th1 T-cell polarization, TCR signaling, and T-cell homeostasis, respectively. Other PBC-associated non-HLA loci are related to B-cell function, TNF signaling, and NF-κB signaling.

Type 1 Diabetes Mellitus (T1DM)

T1DM is a T-cell-mediated autoimmune disorder characterized by the presence of autoantibodies against islet cells. Increasing incidence of T1DM and its potential microvascular and macrovascular complications have shed light on the need for identifying more effective prevention strategies and new treatment targets. To this end, it is essential to enhance our understanding of the pathogenesis of T1DM.

It is a polygenic disorder where the HLA class II genes account for almost half of genetic susceptibility for T1DM. Interestingly, there are loci of these genes that have also been associated with protection from T1DM. Of the so-far-identified non-HLA genes are a variable number of tandem mini-satellite repeats (VNTR) and CTLA-4.

Systemic Lupus Erythematosus (SLE)

It is a chronic autoimmune disease affecting multiple organ systems including the skin, heart, blood, muscle and joints, kidneys, and lungs. As if, SLE is the winner of all fights with the immune system that is not possible unless the immune tolerance against self-components is broken. Activation of innate immune responses and of inflammatory processes along with production of type I interferons and autoantibodies favors pathogenesis of SLE, while mechanisms of clearance of immune complexes such as apoptosis, neutrophil extracellular traps (NET), and nucleic acid sensing are defective. In particular, evidence indicates the multifaceted role of neutrophils in SLE. Lupus neutrophils undergo epigenetic changes causing them to produce higher levels of cytokines that would induce T- and B-cell abnormalities. Also, neutrophils directly contribute to the formation of NET, which is increased in SLE, while clearance of NET materials is impaired in SLE.

As expected, such a complex situation involves contribution by both HLA and non-HLA genes. Several HLA genes including HLA-DRB1, HLA-DQB2, HLA-DQA2, and HLA-DR3 have been associated with susceptibility to SLE and with the autoantibody profile (anti-dsDNA, anti-Ro, and anti-La) in patients with SLE. Genes encoding interferon regulatory factors (IRFs), STAT4, IFIH1, and osteopontin (OPN) contribute to polygenic high IFN signatures, while TREX1, STING, SAMHD1, and TRAP are known to give rise to monogenic high IFN signatures in SLE. Monogenic SLE results from mutation(s) in genes related to classical complement pathway, apoptosis, and antinucleosome antibody production. SLE-associated genes that occur in regulatory regions (e.g., exons, splice sites, introns, and intergenic sites) are TNFAIP3, TNIP1, BLK, ETS1, PRDM1, and IKZF1. Finally, there are SNPs located within the coding region of genes PTPN22 and immunoglobulin-like transcript 3 receptor (ILT3) that have been linked with SLE.

Systemic Sclerosis (SSc)

SSc is considered a complex multisystem disease characterized by a heterogeneous spectrum of clinical manifestations ranging from limited to diffuse cutaneous SSc. Both innate and adaptive immune systems, fibroblasts, and small vessels show abnormal function in SSc.

There is a long list of HLA genes conferring susceptibility to clinical and autoantibody subgroups of SSc. Also, some HLA genes appear to be protective of SSc. Non-HLA genes associated with SSc are known to play a role in innate immunity, interferon signature and inflammation, adaptive immune responses, B- and T-cell proliferation, survival and cytokine production, apoptosis, autophagy, and fibrosis.

Neurological Diseases

There is sufficient evidence to support the immunogenetic basis for some neurological diseases, in particular, multiple sclerosis, Alzheimer's disease, Parkinson's disease, neuromyelitis optica, myasthenia gravis, and amyotrophic lateral sclerosis. Below provides a rapid overview of the immunogenetics of MS as the prototype of such neurological diseases.

Multiple Sclerosis (MS)

It is the most common inflammatory disease of the CNS, which after interfering with normal myelination results in axonal degeneration. The pathologic characteristics of MS are demyelinating lesions, which can be broadly classified according to whether or not autoimmune processes precede demyelination. In this manner, there are lesions arising from T-cell-mediated and T-cell plus antibody-mediated autoimmune encephalomyelitis, while sometimes demyelination is an initial clue resulted from viruses and toxins. Generally, MS is considered a chronic autoimmune disease, which is primarily characterized by activation of CD4+ autoreactive T cells and Th1 T-cell polarization and then by the production of antibodies, complement factors, and CD8+ T cells damaging the CNS tissues.

Studies suggest a strong genetic component for MS; more than 100 genetic loci have been so far identified to confer MS susceptibility. HLA genes totally account for about 10% of the genetic variance of MS. In particular, MHC class II gene HLA-DRB1*1501 is present in about 50% of patients with MS. After that, the IL-7R-α chain gene can explain about 30% of all cases. Also, IL-2RA is another non-HLA gene implicated in MS. Fresh evidence emerged that supports the potential of KIR genes as a risk or protective factor in the immunogenetics of MS. Both CD4+ T cells and NK cells express KIRs. CD4+ T cells that express KIRs are involved in antibody production and NK cells that express KIRs mediate antiviral and antitumoral innate immune responses. Therefore, KIR polymorphisms can affect the individual's risk for MS through the impact they have on antiviral immunity and antibody production. Now there is a big hope for the future of pharmacogenomics of MS when the immunogenetic information may help to predict treatment response, but it is not fulfilled yet.

Infectious Diseases

Human Immunodeficiency Virus (HIV)

There is a variable degree of immunity to HIV. Genetic factors account for about one-fourth of this variation, among which are HLA genes that through interaction with CD8+ cytotoxic T and CD4+ helper T cells help the initiation of anti-HIV adaptive immune responses, thereby conferring resistance to HIV. Studies have shown associations of HLA genes with accelerated disease progression, slow disease progression, protection against infection, reduced viral load levels, and increased susceptibility to infection. In addition, KIR genes have been implicated in resistance to HIV. More interesting is that the development of TB in people with HIV is, at least in part, determined by individual immunogenetic constitution.

Hepatitis B Virus (HBV) and Hepatitis C Virus (HCV) Infection

Infection with HBV or HCV influences the expression of intrahepatic genes, thereby leaving patients liable to chronic liver disease, cirrhosis, and hepatocellular carcinoma. Virus-specific T-cell responses are central to the removal of infected hepatocytes, and therefore, the kinetics of these responses can aid in monitoring clinical recovery. In 1 year, about 2% of patients with chronic HBV infection will experience spontaneous viral clearance, which is characterized by HBV-specific T cells temporary appearing in the peripheral blood. Such an experience does not occur at all in the case of chronic HCV.

Genetic factors contribute to interindividual variation in clinical course of HBV and HCV infection. GWASs have identified HLA and non-HLA that confer susceptibility to persistent HBV infection, progression of disease, and risk of HBV-related hepatocellular carcinoma. Again, both HLA and non-HLA genes have been linked with spontaneous clearance of HCV. Two genes MICA and DEPDC5 demonstrated association with HCV-related hepatocellular carcinoma.

Tuberculosis (TB)

Mycobacterium tuberculosis is recognized by innate immune receptors including TLRs, C-type lectin receptors (CLRs), and NLRs. Upon its recognition, different mechanisms by which immune cells (macrophages and T cells) and cytokines (IL-12, IFN-γ, IL-4, TNF-α, IL-10, IL-6, and TGF-β) can mediate antimycobacterial functions are engaged. Consistently, increasing evidence supports the individual genetic contribution to the control of tuberculosis infection, severe primary tuberculosis, and pulmonary tuberculosis. Genetic risk factors for developing TB include both HLA and non-HLA genes.

Immunogenetics and Immunosenescence

Immunosenescence refers to immune decline with age, which causes increased infection risk and related morbidity and mortality in the elderly. It is characterized by a chronic low-grade inflammation that arises from aberrant innate and adaptive immune responses, playing a role in a myriad of diseases including, but not limited to, atherosclerosis, obesity, type 2 diabetes, osteoporosis, osteoarthritis, neurodegenerative diseases, major depression, and malignancy. In the last four decades, immunosenescence has received great attention from geneticists. However, due to heterogeneous methodology, we are still unable to generalize the findings to all elderly people. In addition to HLA genes, non-HLA genes related to adaptive immunity and to innate immunity appear to contribute to the immunogenetic network of human longevity.

Immunogenetics of Atopic Diseases

The most common chronic diseases afflicting children are asthma, hay fever (allergic rhinitis), and eczema (atopic dermatitis). Due to their strong association with concurrent atopic dermatitis, both asthma and allergic rhinitis can be defined as a type of atopic diseases. In this manner, atopy can be used as an umbrella term that describes a group of diseases, e.g., asthma, allergic rhinitis, food allergy, and urticaria. Atopy is considered a result of a Th2-mediated process, where Th2 cytokines promote the production of IgE by IgE+ memory B cells and plasma cells.

Genome screens show that, in general, atopic diseases are largely heritable (60%) and suggest that shared genes for atopy and other autoimmune diseases lie on the short arm of chromosome 6 where MHC gene are. In particular, the airway epithelium undergoes changes in asthma, and as a result, common genes are likely to cause asthma and other epithelial-based diseases, e.g., Crohn's disease and psoriasis. In addition, genes encoding innate immune receptors and cytokines, which are central to the initiation and progression of allergic responses, contribute to the immunogenetic network of atopy.

Vaccinomics and Adversomics: Immunogenetics and Response to Vaccination

Immune responses to vaccines vary between individuals. The main goal of vaccinomics is to deal with genes that may explain a substantial part of this variation. For example, studies estimate the variation in antibody response to hepatitis B surface antigen (HBsAg), measles virus, mumps virus, and rubella virus to be about 60, 88, 38, and 45% hereditary based. Generally, HLA and non-HLA genes encoding cytokines, cell surface receptors, and TLRs affect immune responses to vaccines, e.g., HBV, smallpox, MMR, and seasonal influenza. Moreover, there is evidence that genetic factors play a role in determining vaccine safety and adverse events, and consequently, it led to the emergence of adversomics. Overall, vaccinomics and adversomics facilitate prediction of vaccine efficacy and safety by using immunogenetic knowledge, and this in itself helps in developing more effective vaccines.

In this manner, *immunogenetics* aims to specify situations in which genetic mutations cause the immune system to be functionally imperfect and secure solutions for them.

WHOLE GENOME SEQUENCING

Whole genome sequencing (also known as WGS, full genome sequencing, complete genome sequencing, or entire genome sequencing) is ostensibly the process of determining the complete DNA sequence of an organism's genome at a single time. This entails sequencing all of an organism's chromosomal DNA as well as DNA contained in the mitochondria and, for plants, in the chloroplast. In practice, genome sequences that are nearly complete are also called whole genome sequences.

Whole genome sequencing has largely been used as a research tool, but is currently being introduced to clinics. In the future of personalized medicine, whole genome sequence data may be an important tool to guide therapeutic intervention. The tool of gene sequencing at SNP level is also used to pinpoint functional variants from association studies and improve the knowledge available to researchers interested in evolutionary biology, and hence may lay the foundation for predicting disease susceptibility and drug response.

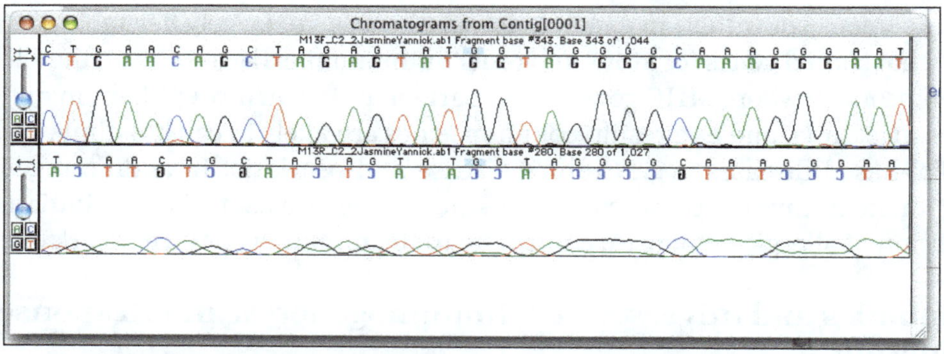

Electropherograms are commonly used to sequence portions of genomes.

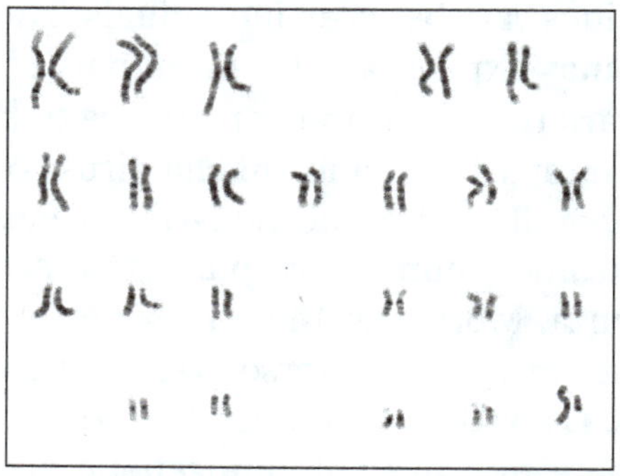

An image of the 46 chromosomes, making up the diploid genome of human male.

Whole genome sequencing should not be confused with DNA profiling, which only determines the likelihood that genetic material came from a particular individual or group, and does not contain additional information on genetic relationships, origin or susceptibility to specific diseases. In addition, whole genome sequencing should not be confused with methods that sequence specific subsets of the genome - such methods include whole exome sequencing (1-2% of the genome) or SNP genotyping (<0.1% of the genome).

As of 2017 there were no complete genomes for any mammals, including humans. Between 4% to 9% of the human genome, mostly satellite DNA, had not been sequenced.

Commercialization

A number of public and private companies are competing to develop a full genome sequencing platform that is commercially robust for both research and clinical use, including Illumina, Knome, Sequenom, 454 Life Sciences, Pacific Biosciences, Complete

Genomics, Helicos Biosciences, GE Global Research (General Electric), Affymetrix, IBM, Intelligent Bio-Systems, Life Technologies, Oxford Nanopore Technologies, and the Beijing Genomics Institute. These companies are heavily financed and backed by venture capitalists, hedge funds, and investment banks.

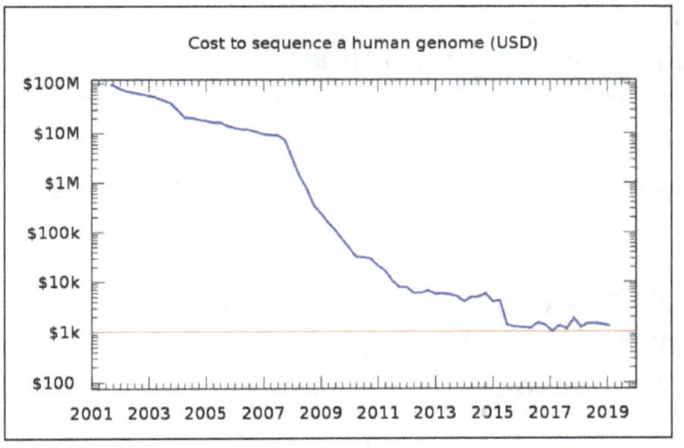

Total cost of sequencing a whole human genome.

A commonly-referenced commercial target for sequencing cost until the late 2010s was $1,000, however, the private companies are working to reach a new target of only $100.

Incentive

In October 2006, the X Prize Foundation, working in collaboration with the J. Craig Venter Science Foundation, established the Archon X Prize for Genomics, intending to award $10 million to "the first team that can build a device and use it to sequence 100 human genomes within 10 days or less, with an accuracy of no more than one error in every 1,000,000 bases sequenced, with sequences accurately covering at least 98% of the genome, and at a recurring cost of no more than $1,000 per genome". The Archon X Prize for Genomics was cancelled in 2013, before its official start date.

Comparison with other Technologies

DNA Microarrays

Full genome sequencing provides information on a genome that is orders of magnitude larger than by DNA arrays, the previous leader in genotyping technology.

For humans, DNA arrays currently provide genotypic information on up to one million genetic variants, while full genome sequencing will provide information on all six billion bases in the human genome, or 3,000 times more data. Because of this, full genome sequencing is considered a disruptive innovation to the DNA array markets as the accuracy of both range from 99.98% to 99.999% (in non-repetitive DNA regions) and their consumables cost of $5000 per 6 billion base pairs is competitive (for some applications) with DNA arrays ($500 per 1 million basepairs).

Applications

Mutation Frequencies

Whole genome sequencing has established the mutation frequency for whole human genomes. The mutation frequency in the whole genome between generations for humans (parent to child) is about 70 new mutations per generation. An even lower level of variation was found comparing whole genome sequencing in blood cells for a pair of monozygotic (identical twins) 100-year-old centenarians. Only 8 somatic differences were found, though somatic variation occurring in less than 20% of blood cells would be undetected.

In the specifically protein coding regions of the human genome, it is estimated that there are about 0.35 mutations that would change the protein sequence between parent/child generations (less than one mutated protein per generation).

In cancer, mutation frequencies are much higher, due to genome instability. This frequency can further depend on patient age, exposure to DNA damaging agents (such as UV-irradiation or components of tobacco smoke) and the activity/inactivity of DNA repair mechanisms. Furthermore, mutation frequency can vary between cancer types: in germline cells, mutation rates occur at approximately 0.023 mutations per megabase, but this number is much higher in breast cancer (1.18-1.66 somatic mutations per Mb), in lung cancer (17.7) or in melanomas (≈33). Since the haploid human genome consists of approximately 3,200 megabases, this translates into about 74 mutations (mostly in noncoding regions) in germline DNA per generation, but 3,776-5,312 somatic mutations per haploid genome in breast cancer, 56,640 in lung cancer and 105,600 in melanomas.

The distribution of somatic mutations across the human genome is very uneven, such that the gene-rich, early-replicating regions receive fewer mutations than gene-poor, late-replicating heterochromatin, likely due to differential DNA repair activity. In particular, the histone modification H3K9me3 is associated with high, and H3K36me3 with low mutation frequencies.

Genome-wide Association Studies

In research, whole-genome sequencing can be used in a Genome-Wide Association Study (GWAS) - a project aiming to determine the genetic variant or variants associated with a disease or some other phenotype.

Diagnostic Use

In 2009, Illumina released its first whole genome sequencers that were approved for clinical as opposed to research-only use and doctors at academic medical centers began quietly using them to try to diagnose what was wrong with people whom standard approaches had failed to help. The price to sequence a genome at that time was US$19,500, which was billed to the patient but usually paid for out of a research grant; one person at that time had applied for reimbursement from their insurance company. For example, one child had needed around 100 surgeries by the time he was three years old, and his doctor turned to whole genome sequencing to determine the problem; it took a team of around 30 people that included 12 bioinformatics experts, three sequencing technicians, five physicians, two genetic counsellors and two ethicists to identify a rare mutation in the XIAP that was causing widespread problems.

Currently available newborn screening for childhood diseases allows detection of rare disorders that can be prevented or better treated by early detection and intervention. Specific genetic tests are also available to determine an etiology when a child's symptoms appear to have a genetic basis. Full genome sequencing, in addition has the potential to reveal a large amount of information (such as carrier status for autosomal recessive disorders, genetic risk factors for complex adult-onset diseases, and other predictive medical and non-medical information) that is currently not completely understood, may not be clinically useful to the child during childhood, and may not necessarily be wanted by the individual upon reaching adulthood.

In 2018, researchers at Rady Children's Institute for Genomic Medicine in San Diego, CA determined that rapid whole-genome sequencing (rWGS) can diagnose genetic disorders in time to change acute medical or surgical management (clinical utility) and improve outcomes in acutely ill infants. The researchers reported a retrospective cohort study of acutely ill inpatient infants in a regional children's hospital from July 2016-March 2017. Forty-two families received rWGS for etiologic diagnosis of genetic disorders. The diagnostic sensitivity of rWGS was 43% (eighteen of 42 infants) and 10% (four of 42 infants) for standard genetic tests ($P = .0005$). The rate of clinical utility of rWGS (31%, thirteen of 42 infants) was significantly greater than for standard genetic tests (2%, one of 42; $P = .0015$). Eleven (26%) infants with diagnostic rWGS avoided morbidity, one had a 43% reduction in likelihood of mortality, and one started palliative care. In six of the eleven infants, the changes in management reduced inpatient cost by $800,000-$2,000,000. These findings replicate a prior study of the clinical utility of rWGS in acutely ill inpatient infants, and demonstrate improved outcomes and net healthcare savings. rWGS merits consideration as a first tier test in this setting.

Due to recent cost reductions whole genome sequencing has become a realistic application in DNA diagnostics. In 2013, the 3Gb-TEST consortium obtained funding from the European Union to prepare the health care system for these innovations in DNA diagnostics. Quality assessment schemes, Health technology assessment and guidelines have to be in place. The 3Gb-TEST consortium has identified the analysis and interpretation of sequence data as the most complicated step in the diagnostic process. At the Consortium meeting in Athens in September 2014, the Consortium coined the word *genotranslation* for this crucial step. This step leads to a so-called *genoreport*. Guidelines are needed to determine the required content of these reports.

Genomes2People (G2P), an initiative of Brigham and Women's Hospital and Harvard Medical School was created in 2011 to examine the integration of genomic sequencing into clinical care of adults and children. G2P's director, Robert C. Green, had previously led the REVEAL study — Risk Evaluation and Education for Alzheimer's Disease – a series of clinical trials exploring patient reactions to the knowledge of their genetic risk for Alzheimer's.

Ethical Concerns

The introduction of whole genome sequencing may have ethical implications. On one hand, genetic testing can potentially diagnose preventable diseases, both in the individual undergoing genetic testing and in their relatives. On the other hand, genetic testing has potential downsides such as genetic discrimination, loss of anonymity, and psychological impacts such as discovery of non-paternity.

Some ethicists insist that the privacy of individuals undergoing genetic testing must be protected. Indeed, privacy issues can be of particular concern when minors undergo genetic testing. This potential use of genome sequencing is highly controversial, as it runs counter to established ethical norms for predictive genetic testing of asymptomatic minors that have been well established in the fields of medical genetics and genetic counseling. The traditional guidelines for genetic testing have been developed over the course of several decades since it first became possible to test for genetic markers associated with disease, prior to the advent of cost-effective, comprehensive genetic screening.

When an individual undergoes whole genome sequencing, they reveal information about not only their own DNA sequences, but also about probable DNA sequences of their close genetic relatives. This information can further reveal useful predictive information about relatives' present and future health risks. Hence, there are important questions about what obligations, if any, are owed to the family members of the individuals who are undergoing genetic testing. In Western/European society, tested individuals are usually encouraged to share important information on any genetic diagnoses with their close relatives, since the importance of the genetic diagnosis for offspring and other close relatives is usually one of the reasons for seeking a genetic testing in the first place. Nevertheless, a major ethical dilemma can develop when the patients refuse to share information on a diagnosis that is made for serious genetic disorder that is highly preventable and where there is a high risk to relatives carrying the same disease mutation. Under such circumstances, the clinician may suspect that the relatives would rather know of the diagnosis and hence the clinician can face a conflict of interest with respect to patient-doctor confidentiality.

Privacy concerns can also arise when whole genome sequencing is used in scientific research studies. Researchers often need to put information on patient's genotypes and phenotypes into public scientific databases, such as locus specific databases. Although only anonymous patient data are submitted to locus specific databases, patients might still be identifiable by their relatives in the case of finding a rare disease or a rare missense mutation. Public discussion around the introduction of advanced forensic techniques (such as advanced familial searching using public DNA ancestry websites and DNA phenotyping approaches) has been limited, disjointed, and unfocused. As forensic genetics and medical genetics converge toward genome sequencing, issues surrounding genetic data become increasingly connected, and additional legal protections may need to be established.

People with Public Genome Sequences

The first nearly complete human genomes sequenced were two Americans of predominantly Northwestern European ancestry in 2007 (J. Craig Venter at 7.5-fold coverage, and James Watson at 7.4-fold). This was followed in 2008 by sequencing of an anonymous Han Chinese man (at 36-fold), a Yoruban man from Nigeria (at 30-fold), and a female caucasian Leukemia patient (at 33 and 14-fold coverage for tumor and normal tissues). Steve Jobs was among the first 20 people to have their whole genome sequenced, reportedly for the cost of $100,000. As of June 2012, there were 69 nearly complete human genomes publicly available. The work was led by Manuel Corpas and the data obtained by direct-to-consumer genetic testing with 23andMe and the Beijing Genomics Institute). This is believed to be the first such Public Genomics dataset for a whole family.

GENETIC COUNSELING

Genetic counseling is the process of advising individuals and families affected by or at risk of genetic disorders to help them understand and adapt to the medical, psychological and familial implications of genetic contributions to disease. The process integrates:

- Interpretation of family and medical histories to assess the chance of disease occurrence or recurrence.
- Education about inheritance, testing, management, prevention, resources.
- Counseling to promote informed choices and adaptation to the risk or condition.

Counseling Session Structure

The goals of genetic counseling are to increase understanding of genetic diseases, discuss disease management options, and explain the risks and benefits of testing. Counseling sessions focus on giving vital, unbiased information and non-directive assistance in the patient's decision-making process. Seymour Kessler, in 1979, first categorized sessions in five phases: an intake phase, an initial contact phase, the encounter phase, the summary phase, and a follow-up phase. The intake and follow-up phases occur outside of the actual counseling session. The initial contact phase is when the counselor and families meet and build rapport. The encounter phase includes dialogue between the counselor and the client about the nature of screening and diagnostic tests. The summary phase provides all the options and decisions available for the next step. If patients wish to go ahead with testing, an appointment is organized and the genetic counselor acts as the person to communicate the results. Result delivery can happen both in person or via phone. Often counselors will call out results to avoid patients having to come back in as results can take weeks to be processed. If further counseling is needed in a more personal setting, or it is determined that additional family members should be tested, a secondary appointment can be made.

Reasons for Testing and Sub-specialties

Detecting Conditions

Diagnostic testing occurs when an individual is showing signs or symptoms associated with a specific condition. Genetic testing can be used to arrive at a definitive diagnosis in order to provide better prognosis as well as medical management and/or treatment options. Testing can reveal conditions that—while debilitating without treatment—can be mild or asymptomatic with early treatment (such as phenylketonuria). Genetic tests are available for a number of genetic conditions, including but not limited to: Down syndrome, sickle cell disease, Tay–Sachs disease, muscular dystrophy. Establishing a genetic diagnosis alerts any other at-risk individuals in the family to consider testing, if appropriate.

Any reproductive risks (e.g. a chance to have a child with the same diagnosis) can also be explored after a diagnosis. Many disorders cannot occur unless both the mother and father pass on their genes, such as cystic fibrosis; this is known as autosomal recessive inheritance. Other autosomal dominant diseases can be inherited from one parent, such as Huntington disease and DiGeorge syndrome. Yet other genetic disorders are caused by an error or mutation occurring during the cell division process (e.g. aneuploidy) and are not hereditary.

Screening tests are often used prior to diagnostic testing, designed to separate people according to a fixed characteristic or property, with the intention of detecting early evidence of disease. For example, if a screening test during a pregnancy (such as maternal blood screening or ultrasound) reveals a risk of a health issue or genetic condition, patients are encouraged to receive genetic counseling to learn additional information regarding the suspected condition and about their options with discussion of management, therapy and treatments available for the conditions, including those after delivery, and those during pregnancy, such as the option to terminate the pregnancy.

Patients may decline additional screening and testing, elect to proceed to diagnostic testing, or pursue further screening tests to refine the risk during the pregnancy.

Presymptomatic or predictive testing occurs when an individual knows of a specific diagnosis (typically adult onset) in their family and has other affected relatives, but they themselves do not manifest any clinical findings at the time when they seek testing. The decision about whether or not to proceed with presymptomatic testing should entail a thoughtful approach and consideration of various medical, reproductive, social, insurance, and financial factors, with no "right" or "wrong" answer. Availability of treatment and medical management options for each specific diagnosis, as well as the genetics and inheritance pattern of the particular condition should be reviewed as inherited conditions can have reduced penetrance.

Insurance and legal issues should also be discussed during genetic counseling. There are laws in the United States such as GINA (Genetic Information Non-discrimination Act) and ACA that provide certain protections against discrimination for individuals with genetic diagnoses.

Adult Genetics

Adult or general genetics clinics serve patients who are diagnosed with genetic conditions that begin to show signs or symptoms in adulthood. Many genetic conditions have varying ages of onset, ranging from an infantile form to an adult form. Genetic counseling can facilitate the decision making process by providing the patient/family with education about the genetic condition as well as the medical management options available to individuals at risk of developing the condition. Having the genetic information of other members of the family opens the door to asking important questions about the pattern of inheritance of specific disease-causing mutations.

ART/Infertility Genetics

Genetic counseling is an integral part of the process for patients utilizing preimplantation genetic testing (PGT), formerly called preimplantation genetic diagnosis. There are three types of PGT and all require in vitro fertilization (IVF) using assisted reproductive technology (ART). PGT-M, for monogenic disorders, involves testing embryos for a specific condition before it is implanted into the mother. This technique is currently being done for disorders with childhood onset, such as Cystic Fibrosis, Tay-Sachs and Muscular Dystrophy, as well as adult-onset conditions, including Huntington's Disease, Hereditary Breast and Ovarian Cancer Syndrome, and Lynch Syndrome. PGT-SR, for structural rearrangements, involves testing embryos to establish a pregnancy unaffected by a structural chromosomal abnormality (translocation). PGT-A, for aneuploidy, was formerly called preimplantation genetic screening, and involved testing embryos to identify any de novo aneuploidy. The indications to carry out PGT-A are: previous aneuploidy in the couple, implantation failure, recurrent miscarriage, severe male factor or advanced maternal age. Finally, PGT seems to be: safe for the embryo, trustable in the diagnosis, more efficient from the reproductive point of view and cost-effective.

Genetic counseling can also involve medical evaluation and clinical work-up for couples with infertility and/or recurrent pregnancy loss, as these histories can be associated with parental chromosome aberrations (such as inversions or translocations) and other genetic conditions.

Cardiovascular Genetics

A rapidly expanding field in genetic counseling is cardiovascular genetics. More than 1 in 200 people have an inherited cardiovascular disease. Hereditary cardiac conditions range from common diseases, such as high cholesterol and coronary artery disease, to rare diseases like Long QT Syndrome, hypertrophic cardiomyopathy, aortic and vascular diseases. Genetic counselors who specialize in cardiovascular disease have developed skills specific to the management of and counseling for cardiovascular disorders. Cardiovascular genetic counselors are also integral in local and national efforts to prevent sudden cardiac death (the leading cause of sudden death in young people) by identifying patients with known or suspected heritable cardiovascular diseases and promoting cascade family screening or testing of at-risk relatives.

Common referral reasons include:

- Cardiomyopathy, including, hypertrophic cardiomyopathy and familial idiopathic dilated cardiomyopathy.
- Arrhythmia, including Long QT syndrome and Brugada syndrome.
- Vascular disease, including Marfan syndrome and aortic aneurysm and/or dissection (<50 years old).
- Other heritable heart defects (e.g. conotruncal defects, left ventricular outflow tract defects).
- Familial hypercholesterolemia.
- Family history of unexplained sudden death.

Guidelines on cardiovascular genetics are published by multiple professional societies.

Hereditary Cancer Genetics

Cancer genetic counselors see individuals with a personal diagnosis and/or family history of cancer or symptoms of an inherited cancer syndrome. Genetic counselors take a family history and assess for hereditary risk, or risk that can be passed down from generation to generation. If indicated, they can coordinate genetic testing, typically via blood or saliva sample, to evaluate for hereditary cancer risk. Personalized medical management and cancer screening recommendations can be provided based on results of genetic testing and/or the family history of cancer. While most cancers are sporadic (not inherited), some are more likely to have a hereditary factor, particularly when occurring at young ages or when clustering in families. These include common cancers such as breast, ovarian, colon and uterine cancers, as well as rare tumor types. General referral indications can include, but are not limited to:

- Personal or family history of cancer with unusually young age of onset (e.g. breast cancer under age 45, colon cancer under age 50).
- Diagnosis of a tumor or cancer with a high likelihood to be hereditary (e.g. male breast cancer, triple negative breast cancer, ovarian cancer, metastatic or high-grade prostate cancer, pancreatic cancer, retinoblastoma, adrenocortical carcinoma, pleuropulmonary blastoma, neuroendocrine tumor, medullary thyroid cancer, pheochromocytoma or paraganglioma).
- Personal or family history of bilateral cancers (e.g., both breasts or both kidneys) or multiple primary tumors in one person.
- Features associated with an inherited cancer syndrome (e.g. more than 10 adenomatous polyps, rare types of gastrointestinal polyps, such as hamartomatous polyps), or specific skin findings like café au lait macules or freckling on the lips.
- Family history of related cancers clustered on the same side of the family (e.g. breast/ovarian, colon/uterine, sarcoma/leukemia/brain).
- Family history of a known inherited cancer syndrome (e.g. hereditary breast and ovarian cancer syndrome, hereditary non-polyposis colorectal cancer, Li-Fraumeni syndrome.
- Individuals of Ashkenazi Jewish ancestry with a personal and/or family history of breast, ovarian, or pancreatic cancer.
- Possible germline (inherited) genetic mutation suggested by tumor profile testing.

Neurogenetics

Genetic counselors specializing in neurogenetics are involved in the care of individuals who have or are at risk to develop conditions affecting the central nervous system (brain and spinal cord) or peripheral nervous system (the nerves that leave the spinal cord and go to other places in the body, such as the feet and hands, skeletal muscles, and internal organs). Effects of these conditions can lead to various impairments some examples of which include cognitive decline, intellectual disability, seizures, uncontrolled movements (e.g. ataxia, chorea), muscle weakness, paralysis, or atrophy. Examples of neurogenetic disorders include:

- Brain malformation syndromes, including lissencephaly and polymicrogyria.
- Brain tumor predisposition syndromes, including Neurofibromatosis 1 and 2.
- Epilepsy (seizures).
- Hereditary motor neuron diseases, including amyotrophic lateral sclerosis (ALS/Lou Gehrig's disease) and spinal muscular atrophy.
- Hereditary neuropathies, including Charcot-Marie-Tooth disease.
- Intellectual disabilities, developmental delays, and autism spectrum disorder.
- Leukodystrophy (hereditary white matter diseases).
- Memory and other cognitive disorders, including Alzheimer disease and frontotemporal dementia.
- Movement disorders, including hereditary ataxia, spastic paraplegia, Huntington disease, and Parkinson disease.
- Neuromuscular disorders, including muscular dystrophies, congenital myopathies, and congenital myasthenic syndromes.

Pediatric Genetics

Pediatric genetic counseling can be indicated for newborns, infants, children and their families. General referral indications can include:

- Birth defects or multiple congenital anomalies (cleft lip/palate, heart defects, spina bifida).
- Intellectual disability of unknown cause, learning disabilities, or autism.
- Sensory impairments (vision, hearing).
- Metabolic disorders (PKU, galactosemia).
- Known/Suspected genetic disorders (e.g., Down syndrome, Cystic Fibrosis, Muscular Dystrophy).

Prenatal Genetics

Prenatal genetics involves services for women either during or prior to a pregnancy.

General indications for referral to genetic counseling in the preconception or prenatal setting may include, but are not limited to:

- Advanced maternal age (35 years old or older at time of delivery).

- Advanced paternal age.
- Current pregnancy with anomalies identified by ultrasound (e.g. increased nuchal translucency measurements).
- Current pregnancy with an abnormal genetic screening test or test result.
- Current pregnancy with risk of or concern for maternal exposures, such as medications, radiation, drugs/alcohol, or infections.
- Consanguineous union (cousins or otherwise blood related).
- Family history of an inherited genetic condition or chromosome abnormality.
- Genetic carrier screening for recessive and/or X-linked diseases.
- History of a previous child with a birth defect, developmental delay, or other genetic condition.
- History of infertility, multiple unexplained miscarriages or cases of unexplained infant deaths.
- Molecular test for single gene disorder.

Prenatal genetic counseling may help with the decision-making process by walking patients through examples of what some people might do in similar situations, and their rationale for choosing that option. Decisions made by patients are affected by factors including timing, accuracy of information provided by tests, and risk and benefits of the tests. This discussion enables patients to place the information and circumstances into the context of their own lives, and in the context of their own values. They may choose to undergo noninvasive screening (e.g. ultrasound, triple screen, cell-free fetal DNA screening) or invasive diagnostic testing (amniocentesis or chorionic villus sampling). Invasive diagnostic tests possess a small risk of miscarriage (1–2%) but provide more definitive results. Testing is offered to provide a definitive answer regarding the presence of a certain genetic condition or chromosomal abnormality.

Psychiatric Genetics

Psychiatric genetic counseling is a sub-Speciality within genetic counseling focused on helping people living with a psychiatric disorder and/or their family members understand both the genetic and environmental factors that contributed to their illness and address associated emotions such as guilt or self-blame. Genetic counselors also discuss strategies to promote recovery and protect mental health and address any questions on chances for recurrence in other family members. While currently there is no single gene solely responsible for causing a psychiatric disorder, there is strong evidence from family, twin studies, and genome-wide-association studies that both multiple genes and environment interact together. Like other areas of genetic counseling, patients at all different stages of life (pediatric, adult, prenatal) can have psychiatric genetic counseling. Since the etiology of psychiatric disorders is complex and not fully understood, the utility of genetic testing is not as clear as it is in Mendelian or single gene disorders. Research has shown that individuals who receive psychiatric genetic counseling have significant increases in feelings of empowerment and self-efficacy after genetic counseling.

Psychiatric genetic counselors can help "dispel mistaken notions about psychiatric disorders, calm needless anxiety, and help those at risk to draw up a rational plan of action based on the best available information".

Psychosocial aspects of Genetic Counseling

There are different approaches to genetic counseling. Some counselors favor a more purely psycho-educational approach while others incorporate more psycho-therapeutic techniques. Genetic counseling is psycho-educational as patients "learn how genetics contributes to their health risks and then process what this means and how it feels."

Psychotherapeutic

"Whether the process of genetic counseling is a form of psychotherapy is up for debate". The relationship between the client and counselor is similar as are the goals of the sessions. As a psychotherapist aims to help his client improve his wellbeing, a genetic counselor also helps his client to address a "situational health threat that similarly threatens client wellbeing". Due to the lack of studies which compare genetic counseling to the practice of psychotherapy, it is hard to say with certainty whether genetic counseling can be "conceptualized as a short-term, applied, specific type of psychotherapy". However, there few existing studies suggest that genetic counseling falls "significantly short of psychotherapeutic counseling" because genetic counseling sessions primarily consist of the distribution of information without much emphasis placed on explaining any long-term impacts to the client.

Genetic Counselors as Support

Genetic counselors provide supportive counseling to families, serve as patient advocates and refer individuals and families to community or state support services. They serve as educators and resource people for other health care professionals and for the general public. Many engage in research activities related to the field of medical genetics and genetic counseling. When communicating increased risk, counselors anticipate the likely distress and prepare patients for the results. Counselors help clients cope with and adapt to the emotional, psychological, medical, social, and economic consequences of the test results.

Each individual considers their family needs, social setting, cultural background, and religious beliefs when interpreting their risk. Clients must evaluate their reasoning to continue with testing at all. Counselors are present to put all the possibilities in perspective and encourage clients to take time to think about their decision. When a risk is found, counselors frequently reassure parents that they were not responsible for the result. An informed choice without pressure or coercion is made when all relevant information has been given and understood.

After counseling for other hereditary conditions, the patient may be presented with the option of having genetic testing. In some circumstances no genetic testing is indicated, other times it may be useful to begin the testing process with an affected family member. The genetic counselor also reviews the advantages and disadvantages of genetic testing with the patient.

Attitudes Toward Genetic Counseling and Genetic Testing

Many studies have examined the attitudes of the lay public toward genetic counseling and genetic testing. Barriers to obtaining genetic counseling include lack of understanding of genetics by both patients and healthcare providers, concerns about cost and insurance, and fears of stigma and/or discrimination.

No simple correlation has been found between the change in technology to the changes in values and beliefs towards genetic testing.

GENETIC DIAGNOSIS OF INTERSEX

Intersex people are born with variations in physical and sex characteristics including those of the chromosomes, gonads, sex hormones, or genitals that, according to the UN Office of the High Commissioner for Human Rights, "do not fit the typical definitions for male or female bodies". Such variations may involve genital ambiguity, and combinations of chromosomal genotype and sexual phenotype other than XY-male and XX-female. Preimplantation genetic diagnosis allows the elimination of embryos and fetuses with intersex traits and thus has an impact on discrimination against intersex people.

Prenatal Hormone Treatment

Currently, prenatal testing and hormone treatment to prevent the physical and behavioral expression of intersex traits is available. It examined the use of "prenatal hormone screening or treatment for the prevention of homosexuality" using research conducted on foetuses with congenital adrenal hyperplasia and other traits.

Alice Dreger, Ellen Feder, and Anne Tamar-Mattis describe how research published by Saroj Nimkarn and Maria New in 2010 constructs "low interest in babies and men – and even interest in what they consider to be men's occupations and games – as "abnormal", and potentially preventable with prenatal dexamethasone". The authors state that "weak and unsupported conclusions" of investigations into the attempted "prevention of benign behavioral sex variations" indicates gaps in the ethical management of clinical research.

In 2012, Hirvikoski and others described a lack of long-term follow-up studies of individuals exposed to prenatal treatment, and the results of a 10-year Swedish study of 43 mothers and children. The authors found evidence of unacceptable side-effects in their study, including neurological consequences. Treatment with dexamethasone was discontinued in Sweden.

GENETIC SCREEN

A genetic screen or mutagenesis screen is an experimental technique used to identify and select for individuals who possess a phenotype of interest in a mutagenized population. Hence a genetic screen is a type of phenotypic screen. Genetic screens can provide important information on gene function as well as the molecular events that underlie a biological process or pathway. While genome projects have identified an extensive inventory of genes in many different organisms, genetic screens can provide valuable insight as to how those genes function.

Basic Screening

Forward genetics (or a forward genetic screen) is an approach used to identify genes (or set of genes) responsible for a particular phenotype of an organism. Reverse genetics (or a reverse genetic screen), on the other hand, analyzes the phenotype of an organism following the disruption of a known gene. In short, forward genetics starts with a phenotype and moves towards identifying the gene(s) responsible, whereas reverse genetics starts with a known gene and assays the effect of its disruption by analyzing the resultant phenotypes. Both forward and reverse genetic screens aim to determine gene function.

Successful forward genetic screens often have two key components. The first is a defined genetic background of the organism being used and the second is a simple yet constant experimental procedure to identify mutants of interest. Defined genetic backgrounds allow researchers to identify and locate affected genes in mutant individuals with greater efficiency. A simplified screening method is beneficial because it allows for a larger number of individuals to be screened, thereby increasing the probability of generating and identifying mutants of interest.

Since natural allelic mutations are rare, prior to screening geneticists often mutagenize a population of individuals by exposing them to a known mutagen, such as a chemical or radiation, thereby generating a much higher frequency of chromosomal mutations. In some organisms mutagens can be useful for performing saturation screens. Saturation screens are used to uncover all genes involved in a particular phenotype of an organism or species. The screen is carried out by mapping mutants of a biological process until no new genes/gene mutations can be found. Christiane Nüsslein-Volhard and Eric Wieschaus were the first individuals to perform this type of screening procedure.

Screening Variations

Many screening variations have been devised to elucidate a gene that leads to a mutant phenotype of interest.

Enhancer

An enhancer screen begins with a mutant individual that has an affected process of interest with a known gene mutation. The screen can then be used to identify additional genes or gene mutations that play a role in that biological or physiological process. A genetic enhancer screen identifies mutations which enhance a phenotype of interest in an already mutant individual. The phenotype of the double mutant (individual with both the enhancer and original background mutation) is more prominent than either of the single mutant phenotypes. The enhancement must surpass the expected phenotypes of the two mutations on their own, and therefore each mutation may be considered an enhancer of the other. Isolating enhancer mutants can lead to the identification of interacting genes or genes which act redundantly with respect to one another.

Suppressor

A suppressor screen is used to identify suppressor mutations which alleviate or revert the phenotype of the original mutation, in a process defined as synthetic viability. Suppressor mutations can be described as second mutations at a site on the chromosome distinct from the mutation under study, which suppress the phenotype of the original mutation. If the mutation is in the same gene as the original mutation it is known as intragenic suppression, whereas a mutation located in a different gene is known as extragenic suppression or intergenic suppression. Suppressor mutations are extremely useful to define the functions of biochemical pathways within a cell and the relationships between different biochemical pathways.

Temperature Sensitive

A temperature sensitive screen involves performing temperature shifts to enhance a mutant phenotype. A population grown at low temperature would have a normal phenotype; however, the mutation in the particular gene would make it unstable at a higher temperature. A screen for temperature sensitivity in fruit flies, for example, might involve raising the temperature in the cage until some flies faint, then opening a portal to let the others escape. Individuals selected in a screen are liable to carry an unusual version of a gene involved in the phenotype of interest. An advantage of alleles found in this type of screen is that the mutant phenotype is conditional and can be activated by simply raising the temperature. A null mutation in such a gene may be lethal to the embryo and such mutants would be missed in a basic screen. A famous temperature sensitive screen was carried out independently by Lee Hartwell and Paul Nurse to identify mutants defective in cell cycle in *S. cerevisiae* and *S. pombe*, respectively.

Mapping Mutants

By the classical genetics approach, a researcher would then locate (map) the gene on its chromosome by crossbreeding with individuals that carry other unusual traits and collecting statistics on how frequently the two traits are inherited together. Classical geneticists would have used phenotypic traits to map the new mutant alleles. With the advent of genomic sequences for model systems such as *Drosophila melanogaster, Arabidopsis thaliana* and *C. elegans* many single nucleotide polymorphisms (SNPs) have now been identified that can be used as traits for mapping. In fact, the Heidelberg screen, which was developed in 1980 by Nüsslein-Volhard and Wieschaus cleared the way for future scientists in this field. SNPs are the preferred traits for mapping since they are very frequent, on the order of one difference per 1000 base pairs, between different varieties of organism. Mutagens such as random DNA insertions by transformation or active transposons can also be used to generate new mutants. These techniques have the advantage of tagging the new alleles with a known molecular (DNA) marker that can facilitate the rapid identification of the gene.

Positional Cloning

Positional cloning is a method of gene identification in which a gene for a specific phenotype is identified only by its approximate chromosomal location (but not the function); this is known as the candidate region. Initially, the candidate region can be defined using techniques such as linkage analysis, and positional cloning is then used to narrow the candidate region until the gene and its mutations are found. Positional cloning typically involves the isolation of partially overlapping DNA segments from genomic libraries to progress along the chromosome toward a specific gene. During the course of positional cloning, one needs to determine whether the DNA segment currently under consideration is part of the gene.

Tests used for this purpose include cross-species hybridization, identification of unmethylated CpG islands, exon trapping, direct cDNA selection, computer analysis of DNA sequence, mutation screening in affected individuals, and tests of gene expression. For genomes in which the regions of genetic polymorphisms are known, positional cloning involves identifying polymorphisms that flank the mutation. This process requires that DNA fragments from the closest known genetic marker are progressively cloned and sequenced, getting closer to the mutant allele with each new clone. This process produces a contig map of the locus and is known as chromosome walking. With the completion of genome sequencing projects such as the Human Genome Project, modern positional cloning can use ready-made contigs from the genome sequence databases directly.

For each new DNA clone a polymorphism is identified and tested in the mapping population for its recombination frequency compared to the mutant phenotype. When the DNA clone is at or close to the mutant allele, the recombination frequency should be close to zero. If the chromosome walk proceeds through the mutant allele, the new polymorphisms will start to show increase in recombination frequency compared to the mutant phenotype. Depending on the size of the mapping population, the mutant allele can be narrowed down to a small region (<30 Kb). Sequence comparison between wild type and mutant DNA in that region is then required to locate the DNA mutation that causes the phenotypic difference.

Modern positional cloning can more directly extract information from genomic sequencing projects and existing data by analyzing the genes in the candidate region. Potential disease genes from the candidate region can then be prioritized, potentially reducing the amount of work involved. Genes with expression patterns consistent with the disease phenotype, showing a (putative) function related to the phenotype, or homologous to another gene linked to the phenotype are all priority candidates. Generalization of positional cloning techniques in this manner is also known as positional gene discovery.

Positional cloning is an effective method to isolate disease genes in an unbiased manner, and it has been used to identify disease genes for Duchenne muscular dystrophy, Huntington's disease, and cystic fibrosis. However, complications in the analysis arise if the disease exhibits locus heterogeneity.

GENETIC TESTING

Genetic testing is a type of medical test that identifies changes in chromosomes, genes, or proteins. The results of a genetic test can confirm or rule out a suspected genetic condition or help determine a person's chance of developing or passing on a genetic disorder. More than 1,000 genetic tests are currently in use, and more are being developed.

Several methods can be used for genetic testing:

- Molecular genetic tests (or gene tests) study single genes or short lengths of DNA to identify variations or mutations that lead to a genetic disorder.

- Chromosomal genetic tests analyze whole chromosomes or long lengths of DNA to see if there are large genetic changes, such as an extra copy of a chromosome, that cause a genetic condition.
- Biochemical genetic tests study the amount or activity level of proteins; abnormalities in either can indicate changes to the DNA that result in a genetic disorder.

Genetic testing is voluntary. Because testing has benefits as well as limitations and risks, the decision about whether to be tested is a personal and complex one. A geneticist or genetic counselor can help by providing information about the pros and cons of the test and discussing the social and emotional aspects of testing.

Types of Genetic Tests

Newborn Screening

Newborn screening is used just after birth to identify genetic disorders that can be treated early in life. Millions of babies are tested each year in the United States. All states currently test infants for phenylketonuria (a genetic disorder that causes intellectual disability if left untreated) and congenital hypothyroidism (a disorder of the thyroid gland). Most states also test for other genetic disorders.

Diagnostic Testing

Diagnostic testing is used to identify or rule out a specific genetic or chromosomal condition. In many cases, genetic testing is used to confirm a diagnosis when a particular condition is suspected based on physical signs and symptoms. Diagnostic testing can be performed before birth or at any time during a person's life, but is not available for all genes or all genetic conditions. The results of a diagnostic test can influence a person's choices about health care and the management of the disorder.

Carrier Testing

Carrier testing is used to identify people who carry one copy of a gene mutation that, when present in two copies, causes a genetic disorder. This type of testing is offered to individuals who have a family history of a genetic disorder and to people in certain ethnic groups with an increased risk of specific genetic conditions. If both parents are tested, the test can provide information about a couple's risk of having a child with a genetic condition.

Prenatal Testing

Prenatal testing is used to detect changes in a fetus's genes or chromosomes before birth. This type of testing is offered during pregnancy if there is an increased risk that the baby will have a genetic or chromosomal disorder. In some cases, prenatal testing can lessen a couple's uncertainty or help them make decisions about a pregnancy. It cannot identify all possible inherited disorders and birth defects, however.

Preimplantation Testing

Preimplantation testing, also called preimplantation genetic diagnosis (PGD), is a specialized technique that can reduce the risk of having a child with a particular genetic or chromosomal disorder. It is used to detect genetic changes in embryos that were created using assisted reproductive techniques such as in-vitro fertilization. In-vitro fertilization involves removing egg cells from a woman's ovaries and fertilizing them with sperm cells outside the body. To perform preimplantation testing, a small number of cells are taken from these embryos and tested for certain genetic changes. Only embryos without these changes are implanted in the uterus to initiate a pregnancy.

Predictive and Presymptomatic Testing

Predictive and presymptomatic types of testing are used to detect gene mutations associated with disorders that appear after birth, often later in life. These tests can be helpful to people who have a family member with a genetic disorder, but who have no features of the disorder themselves at the time of testing. Predictive testing can identify mutations that increase a person's risk of developing disorders with a genetic basis, such as certain types of cancer. Presymptomatic testing can determine whether a person will develop a genetic disorder, such as hereditary hemochromatosis (an iron overload disorder), before any signs or symptoms appear. The results of predictive and presymptomatic testing can provide information about a person's risk of developing a specific disorder and help with making decisions about medical care.

Forensic Testing

Forensic testing uses DNA sequences to identify an individual for legal purposes. Unlike the tests described above, forensic testing is not used to detect gene mutations associated with disease. This type of testing can identify crime or catastrophe victims, rule out or implicate a crime suspect, or establish biological relationships between people (for example, paternity).

References

- American-board-of-medical-genetics-and-genomics, contact-an-abms-member-board, member-boards: abms.org, Retrieved 13 July, 2020
- Strachnan T, Read AP (2004). Human Molecular Genetics (3rd ed.). Garland Publishing. p. 616. ISBN 978-0-8153-4184-0
- Genetic-engineering, science: britannica.com, Retrieved 27 June, 2020
- Dinculescu A, Glushakova L, Min SH, Hauswirth WW (June 2005). "Adeno-associated

virus-vectored gene therapy for retinal disease". Hum. Gene Ther. 16(6): 649–63. doi:10.1089/hum.2005.16.649. PMID 15960597

- Introductory-chapter-immunogenetics, immunogenetics, books: intechopen.com, Retrieved 17 April, 2020
- Sibley, Lisa (2009-02-08). "Complete Genomics brings radical reduction in cost". Silicon Valley / San Jose Business Journal. Sanjose.bizjournals.com. Retrieved 2009-02-23

5

Molecular Medicine

The field of medicine that uses physical, chemical, bioinformatics and medical techniques for describing molecular structures, identifying molecular and genetic errors of disease, and developing molecular mechanisms. All the aspects related to molecular medicine have been carefully analyzed in this chapter.

Molecular medicine is a broad field, where physical, chemical, biological and medical techniques are used to describe molecular structures and mechanisms, identify fundamental molecular and genetic errors of disease, and to develop molecular interventions to correct them. The molecular medicine perspective emphasizes cellular and molecular phenomena and interventions rather than the previous conceptual and observational focus on patients and their organs.

MOLECULAR DIAGNOSTICS

Molecular diagnostics is a rapidly growing discipline in laboratory medicine, with new methods and new applications continually becoming available and improvements being made. Molecular techniques are routinely used in the diagnosis and management of many infectious diseases and are increasingly being applied to other areas such as cancer detection. A wide variety of nucleic acid detection methods are available, including amplified and non-amplified techniques. Many assays are commercially available and approved by the U.S. Food and Drug Administration (FDA), allowing laboratories to implement them relatively easily. Improvements to molecular assays include simplified formats (extraction, amplification, and detection in a single device) and automated extraction platforms that allow large numbers of samples to be processed at one time. Qualitative as well as quantitative molecular techniques are commercially available, and some laboratories develop their own home-brew assays. However, several fundamental features of molecular testing require laboratories to be alert for problems with testing that may give rise to erroneous results. The more common errors that can occur with molecular testing are discussed in this chapter. As with other sections of the laboratory, a strong program of quality control and quality assurance is necessary for detection of problems, monitoring of errors, and implementing methods to improve quality if systematic errors are occurring in the molecular laboratory. Although analytical errors can occur with molecular testing, good-quality control programs mean that errors are kept to a minimum.

MOLECULAR PATHOLOGICAL EPIDEMIOLOGY

Molecular pathological epidemiology (MPE, also molecular pathologic epidemiology) is a discipline combining epidemiology and pathology. It is defined as "epidemiology of molecular pathology and heterogeneity of disease". Pathology and epidemiology share the same goal of elucidating etiology of disease, and MPE aims to achieve this goal at molecular, individual and population levels. Typically, MPE utilizes tissue pathology resources and data within existing epidemiology studies. Molecular epidemiology broadly encompasses MPE and conventional-type molecular epidemiology with the use of traditional disease designation systems.

Disease Process

Data from The Cancer Genome Atlas projects indicate that disease evolution is an inherently heterogeneous process. Each patient has a unique disease process ("the unique disease principle"), considering the uniqueness of the exposome and its unique influence on molecular pathologic process. This concept has been adopted in clinical medicine along with precision medicine and personalized medicine.

Methodology

In MPE, investigators dissect interrelationships between exposures (e.g., environmental, dietary, lifestyle and genetic factors); alterations in cellular or extracellular molecules (disease molecular signatures); and disease evolution and progression. Investigators can analyze genome, methylome, epigenome, metabolome, transcriptome, proteome, microbiome, immunity and interactome. A putative risk factor can be linked to specific molecular signatures.

MPE research enables identification of a new biomarker for potential clinical utility, using large-scale population-based data (e.g., PIK3CA mutation in colorectal cancer to select patients for aspirin therapy). The MPE approach can be used following a genome-wide association study (GWAS), termed "GWAS-MPE approach". Detailed disease endpoint phenotyping can be conducted by means of molecular pathology or surrogate histopathology or immunohistochemistry analysis of diseased tissues and cells within GWAS.

As an alternative approach, potential risk variants identified by GWAS can be examined in combination with molecular pathology analysis on diseased tissues. This GWAS-MPE approach can give not only more precise effect estimates, even larger effects, for specific molecular subtypes of the disease, but also insights into pathogenesis by linking genetic variants to molecular pathologic signatures of disease. Since molecular diagnostics is becoming routine clinical practice, molecular pathology data can aid epidemiologic research.

MOLECULAR PATHOLOGY

Molecular pathology is an emerging discipline within pathology which is focused in the study and diagnosis of disease through the examination of molecules within organs, tissues or bodily fluids. Molecular pathology shares some aspects of practice with both anatomic pathology and clinical pathology, molecular biology, biochemistry, proteomics and genetics, and is sometimes considered a "crossover" discipline. It is multi-disciplinary in nature and focuses mainly on the sub-microscopic aspects of disease. A key consideration is that more accurate diagnosis is possible when the diagnosis is based on both the morphologic changes in tissues (traditional anatomic pathology) and on molecular testing.

It is a scientific discipline that encompasses the development of molecular and genetic approaches to the diagnosis and classification of human diseases, the design and validation of predictive biomarkers for treatment response and disease progression, the susceptibility of individuals of different genetic constitution to develop disorders.

Molecular pathology is commonly used in diagnosis of cancer and infectious diseases. Techniques are numerous but include quantitative polymerase chain reaction (qPCR), multiplex PCR, DNA microarray, in situ hybridization, in situ RNA sequencing, DNA sequencing, antibody based immunofluorescence tissue assays, molecular profiling of pathogens, and analysis of bacterial genes for antimicrobial resistance.

Integration of "molecular pathology" and "epidemiology" led to an interdisciplinary field, termed "molecular pathological epidemiology" (MPE), which represents integrative molecular biologic and population health science.

Molecular Pathology of Prostate Cancer

Rostate cancer is the second leading cause of cancer deaths in men. It is not invariably lethal, however, and is a heterogeneous disease ranging from asymptomatic to a rapidly fatal systemic malignancy. The prevalence of prostate cancer is so high that it could be considered a normal age related phenomenon: in a Spanish study examining white Mediterranean men, 33% of men in their 8th decade had evidence of prostate cancer at necropsy and died with the disease, but not from it. Data from American postmortem studies show an even higher prostate cancer prevalence rate.1 In recent years, there have been large increases in the five year survival rates for prostate cancer, with a five year relative age standardised survival rate of 65% in England and Wales for the years 1996–9. This was the third highest survival rate of all cancers over this time period, with only testicular cancer and melanoma having better survival rates, and was around 11% higher than that for patients diagnosed during 1991.

"The prevalence of prostate cancer is so high that it could be considered a normal age related phenomenon".

Unfortunately, this improvement does not reflect better treatment for prostate cancer. It largely reflects an increasing number of men being diagnosed with very early stage prostate cancer as a result of the widespread use of prostate specific antigen (PSA) testing. PSA is a protein produced by both normal and cancerous prostate cells and a high PSA value can be a sign of cancer. Most men diagnosed at a very early stage will die with prostate cancer but not from it; therefore, the survival rate has increased. The American Cancer Society recommends in its prostate cancer screening guidelines that men should be informed of what is known and what is uncertain about the benefits and limitations of early detection of prostate cancer, so that they can make an informed decision about testing. Therefore, the early diagnosis of prostate cancer through screening creates difficulties in predicting the outcome of individual patients. The difficulty is in distinguishing between clinically indolent prostate cancers, which will be asymptomatic, and aggressive prostate cancers with the potential to kill the patient. Gleason grading on histopathological examination is the best prognostic indicator to date in prostate cancer; however, interobserver variation can occur, grading on biopsies may not correlate with the prostatectomy specimen because of sampling problems, and cases of morphologically identical prostate cancer can behave differently.

This is an exciting era with the emergence of new investigative tools such as DNA microarray technology and the application of the field of proteomics to the study of human cancers. Knowledge of genetic changes underlying the initiation, development, and progression of prostate cancer is accumulating rapidly. With increasing knowledge it may be possible to distinguish indolent from aggressive prostate tumours by molecular fingerprinting. A clinical application of this knowledge would be that radical treatment and its associated morbidity could be avoided in prostate cancers that are unlikely to progress. Resources and radical treatment could be focused on prostate cancers with poor prognostic indicators. In this review, we shall discuss the most consistently reported molecular pathological findings in prostate cancer, together with new concepts and technologies.

Hereditary Prostate Cancer

Prostate cancer can be divided epidemiologically into hereditary and sporadic forms,3 but it is not possible to distinguish these two groups at a molecular level. Highly penetrant inherited genes conferring the prostate cancer phenotype have not been identified.

Linkage studies using genetic markers to search for chromosomal regions that show excessive sharing of inherited alleles in cancer affected families have been helpful in identifying important cancer susceptibility genes in other cancers. However, similar studies using families prone to prostate cancer have not yielded the same success.

"The failure to identify highly penetrant genes in hereditary prostate cancer may result from the fact that multiple genes with a small to moderate effect are involved in hereditary prostate carcinogenesis".

Although possible inherited prostate cancer susceptibility genes have been identified such as the ELAC2, RNASEL, MSR1, NSB1, and CHEK2 genes in some families, the proportion of cases of hereditary prostate cancer attributable to germline mutations in these loci is small. Many studies have not supported the role of these genes in hereditary prostate cancer. Mutations of these candidate genes have also been identified in sporadic prostate cancer. Because prostate cancer is a common cancer, it may be difficult to distinguish clustering of sporadic prostate cancer within families from true hereditary prostate cancer. This difficulty may have hindered research into hereditary prostate cancer to date. Alternatively, the failure to identify highly penetrant genes in hereditary prostate cancer may result from the fact that multiple genes with a small to moderate effect are involved in hereditary prostate carcinogenesis. The risk of disease in the presence of a susceptibility gene might be substantially increased only in the appropriate genetic, dietary, and environmental background. We will briefly outline the most important hereditary prostate cancer susceptibility genes identified to date.

ELAC2

ELAC2 was the first possible hereditary prostate cancer gene to be identified. The function of ELAC2 is not definitively known and it has been proposed as a metal dependent hydrolase. An association of ELAC2 genotypes with familial prostate cancer has been reported.5 However, multiple large subsequent studies have not provided confirmatory evidence of this association.

Overall, it appears that if ELAC2 plays a role in prostate cancer it is a relatively minor role.

Host Response to Infection Genes

RNASEL

RNASEL is a ribonuclease that degrades viral and cellular RNA and can produce apoptosis on viral infection. Mutations in the RNASEL gene have been identified in familial and sporadic prostate cancer in many studies,8–12 although other studies have not supported these findings.

Overall, there is strong support for the notion that RNASEL is the most important hereditary prostate cancer gene identified to date.

MSR1

MSR1 encodes a macrophage scavenger receptor responsible for cellular uptake of molecules, including bacterial cell wall products. The importance of MSR1 as a prostate cancer susceptibility gene in hereditary prostate cancer is controversial. Germline MSR1 mutations have been linked to prostate cancer in some families with prostate cancer and in sporadic prostate cancer. However, a recent report, which investigated 163 families with familial prostate cancer, did not provide confirmatory evidence of the role of MSR1 in familial prostate cancer.

Mutations of these host response to infection genes may increase the risk of prostate cancer by predisposing to chronic inflammation as a result of failure of viral RNA and bacterial degradation. There is accumulating knowledge supporting the role of inflammation in prostate cancer, which we will refer to again later in the article.

Cell Cycle Checkpoint Genes

NBS1

The rare human genetic disorder, Nijmegen breakage syndrome, is characterised by radiosensitivity, immunodeficiency, chromosomal instability, and an increased risk for cancer of the lymphatic system. The NBS1 gene, which is involved in this human genetic disorder, encodes a protein, nibrin, involved in the processing/repair of DNA double strand breaks and in cell cycle checkpoints.18 Mutations in the gene for the Nijmegen breakage syndrome (NBS1) have been identified in both sporadic and familial cases of prostate cancer and are associated with a small increased risk of prostate cancer.

CHEK2

The CHEK2 gene is an upstream regulator of p53 in the DNA damage signalling pathway. CHEK2 mutations have been identified in both sporadic and familial cases of prostate cancer and are associated with a small increased risk of prostate cancer.

NBS1 and CHEK2 genes have only recently been identified as possible prostate cancer susceptibility genes. ELAC2 was the first hereditary prostate cancer susceptibility gene identified and subsequent studies have not provided confirmatory evidence of its role in prostate cancer. Therefore, it is not possible to comment on the importance of these two genes in hereditary prostate cancer until additional confirmatory studies have been performed.

The study of hereditary prostate cancer genes is in its infancy and the challenge for the future will be to detect genes with small to moderate effects. Advances in statistical methods to amplify signals from susceptibility genes in the presence of heterogeneous factors are required to decipher the genetics and molecular pathology of hereditary prostate cancer.

Table: Hereditary prostate cancer genes.

Gene	Chromosomal locus	Putative function	Status in prostate cancer
ELAC2	17p	Metal dependent hydrolase.	Unknown
RNASEL	1q	Ribonuclease that degrades viral and cellular RNA and can produce apoptosis on viral infection.	Deleted
MSR1	8p	Encodes a macrophage scavenger receptor responsible for cellular uptake of molecules, including bacterial cell wall products.	Deleted
NBSI	5p	Encodes a protein, nibrin, involved in the processing/repair of DNA double strand breaks and in cell cycle checkpoints.	Deleted
CHEK2	22q	Upstream regulator of p53 in the DNA damage signalling pathway.	Deleted

Sporadic Prostate Cancer

Polymorphisms Associated with Increased Prostate Cancer Risk

A polymorphism is a genetic variant that appears in at least 1% of the population. These common genetic polymorphisms probably have small relative risks, yet large population attributable risks because of their high frequencies.

TLR4

TLR4 encodes a receptor that is a central player in the signalling pathways of the innate immune response to infection by Gram negative bacteria. A TLR4 sequence polymorphism is associated with a small increased risk of prostate cancer. This is in keeping with the current hypothesis of inflammation having a role in prostate carcinogenesis.

CDKN1B

The loss of cell cycle control is thought to be an important mechanism in the promotion of carcinogenesis. CDKN1B belongs to the Cip/Kip family and functions as an important cell cycle gatekeeper.

A recent study has revealed a significant association between a single nucleotide polymorphism of CDKN1B and prostate cancer. There is also an association between CDKN1B and another Cip/Kip family member, CDKN1A, and advanced prostate cancer.

Androgen Receptor (AR)

Growth of prostate cells depends on androgens. Genes that encode products that play a role in inducing androgen stimulation of the prostate gland are very important. The androgen receptor (AR) is currently a therapeutic target for the treatment of prostate cancer. Other genes involved in androgen stimulation of the prostate such as SRD5A2 and CYP17 also hold potential as future therapeutic targets.

Table: Polymorphisms associated with increased prostate cancer risk.

Gene	Chromosomal locus	Function
TLR4	9q	Encodes a receptor that is a central player in the signalling pathways of the innate immune response to infection by Gram negative bacteria.
CDKN1B (p27)	12p	Belongs to the Cip/Kip family and functions as an important cell cycle gatekeeper.
AR	Xq	May cause activation of androgen dependent genes.
CYP17	10q	Enzyme responsible for the biosynthesis of testosterone.
SRD5A2	2p	Converts testosterone to the more potent dihydrotestosterone.
AR, androgen receptor		

The AR contains polymorphic polyglutamine (CAG)n trinucleotide repeats. It has been reported in the past that shortening of these repeats is associated with increased prostate cancer risk.

Short CAG length has also been correlated with high grade, high stage, metastatic, and fatal prostate cancers. A hypothesis that has been proposed for the influence of the short CAG repeat on prostate carcinogenesis is that because of its role in AR function it causes an increase in activation of androgen dependent genes.

CYP17

CYP17 encodes cytochrome P-450c17a, an enzyme responsible for the biosynthesis of testosterone. A variant CYP17 allele is associated with both hereditary and sporadic prostate cancer. This allele is hypothesised to increase the rate of gene transcription, increase androgen production, and thereby increase the risk of prostate cancer.

SRD5A2

SRD5A2 encodes the predominant isozyme of 5a-reductase in the prostate, an enzyme that converts testosterone to the more potent dihydrotestosterone. The alleles that encode enzymes with increased activity have been associated with an increased risk of prostate cancer and with a poor prognosis for men with prostate cancer.

Polymorphisms Associated with Advanced Sporadic Prostate Cancer

Vitamin D Receptor

Physiological concentrations of vitamin D promote the differentiation and growth arrest of prostate cancer cells in vitro. The precise mechanism through which vitamin D mediates this effect is unknown, although it is probably through its effect on cell growth proteins. Allelic differences in the vitamin D receptor (VDR) gene result in variation in VDR activity. VDR alleles have been significantly associated with prostate cancer and this association was stronger with advanced prostate cancers.Polymorphisms in CDKN1A and CDKN1B are associated with advanced prostate cancer. In addition, polymorphic variants of several other genes have been proposed as possible contributors to the risk of prostate cancer.

Table: Polymorphisms associated with advanced prostate cancer risk.

Gene	Chromosomal locus	Function
Vitamin D receptor	13q	Promotes the differentiation and growth arres.
CDKN1A	6p	Belong to the Cip/Kip family and functions as an important cell cycle gatekeeper.
CDKN1B	12p	

Somatic Genetic Changes

The number of genetic loci involved in prostate carcinogenesis is large and the mechanisms are complex and not fully understood. The most commonly reported chromosomal abnormalities in sporadic prostate cancer, together with the putative genes involved at these chromosomal sites.

Although these are the most common areas of chromosomal loss and gain, prostate carcinogenesis is complex and multiple genes from other chromosomal loci are also thought to be involved.

Tumour Suppressor Genes and Loss of Heterozygosity

Tumour suppressor genes are probably involved in the prostate carcinogenesis pathway. Loss of tumour suppressor genes was initially proposed to occur via loss of function of two alleles (the "two hit hypothesis") by mutation or deletion. This model has been revised to include epigenetic modification by (a) inactivation of one or both alleles by DNA methylation of CpG sites in gene promoters, (b) function heritably downregulated, or (c) otherwise compromised in a clonal fashion. The change can be by mutation, methylation of the promoter, or by some other modification of the protein product, and must be coupled with evidence that the normal (wild-type) gene does suppress growth of tumour cells.

Glutathione S-transferase Gene

The glutathione S-transferase (GSTP1) gene is emerging as one of the most important tumour suppressor genes in prostate cancer. GSTP1 can detoxify environmental electrophilic carcinogens and oxidants and may play a genome caretaker role by preventing oxidant and electrophilic DNA damage. GSTP1 has been shown to be inactivated by hypermethylation of the promoter region in prostate tumours. Hypermethylation of GSTP1 is the most common (90%) reported epigenetic alteration in prostate cancer. It occurs early in cancer progression and is a promising marker for detecting organ confined disease. The quantitation of GSTP1 hypermethylation can accurately detect the presence of cancer even in small, limited tissue samples. It is a promising diagnostic marker that could possibly be used as an adjunct to tissue biopsy as part of prostate cancer screening.

Aberrant DNA methylation patterns may be the earliest somatic genome changes in prostate cancer. A recent study found that CpG islands were hypermethylated in. 85% of prostate cancers and cancer cell lines but not in normal prostate cells and tissues. CpG island hypermethylation patterns in prostate cancer metastases were very similar to the primary prostate cancers and tended to show greater differences between cases than between anatomical sites of metastasis.

PTEN

PTEN (phosphatase and tensin homologue) is an important tumour suppressor gene in prostate cancer and influences the concentrations of CDKN1B another important tumour suppressor gene. The PTEN gene encodes a phosphatase active against both protein and lipid substrates and is a common target for somatic alteration during the progression of prostate cancer. PTEN is present in normal epithelial cells and in cells in prostatic intraepithelial neoplasia (PIN). In prostate cancers, concentrations of PTEN are often reduced, particularly in cancers of high grade or stage. In prostate cancers that do contain PTEN, there is considerable heterogeneity in concentrations, with regions devoid of PTEN being described. The mechanism by which PTEN might act as a tumour suppressor gene in the prostate may involve inhibition of the phosphatidylinositol 39-kinase–protein kinase B (P13K–Akt) signalling pathway, which is essential for cell cycle progression and cell survival.

CDKN1B (p27)

CDKN1B (p27) is an important tumour suppressor gene in prostate cancer. Reduced concentrations of p27, a cyclin dependent kinase inhibitor encoded by the CDKN1B gene, are common in prostate cancers, particularly in those with a poor prognosis. The somatic loss of DNA sequences at 12p12–3, encompassing CDKN1B, has been described in 23% of localised prostate cancers, 30% of prostate cancer metastases in regional lymph nodes, and 47% of distant prostate cancer metastases.61 Concentrations of p27 are suppressed by the P13K–Akt signalling pathway.58 By inhibiting P13K–Akt, PTEN can increase the concentration of CDKN1B mRNA and p27 protein. For this reason, low p27 concentrations may be as much a result of the loss of PTEN function as of CDKN1B alterations.

NKX3.1

Loss of 8p appears to be an early event in prostate cancer and the most promising candidate tumour suppressor gene at this site is NKX3.1, which encodes a prostate specific homeobox gene that is probably essential for normal prostate development. NKX3.1 binds to DNA and represses expression of the PSA gene.63 Loss of NKX3.1 expression appears to be related to the progression of prostate cancer. One study found that NKX3.1 was absent in 20% of PIN lesions, 6% of low stage prostate cancers, 22% of high stage prostate cancers, 34% of androgen independent prostate cancers, and 78% of prostate cancer metastases.42 The loss of this gene is of particular interest because when present it represses expression of the PSA gene, and the loss of NKX3.1 may be involved in the increasing concentrations of PSA seen with prostate cancer progression.

Table: The most commonly described areas of chromosomal loss and gain in prostate cancer.

Chromosome locus	Putative genes	Normal function of gene	Status of gene in prostate cancer
7p	EGFR	Growth factor	Amplified
7q	CAV 1	Structural protein of caveolae membranes in fibroblasts and endothelia	Amplified
8p	MSR	Encodes a macrophage scavenger receptor responsible for cellular uptake of molecules including bacterial cell wall products	Deleted
8p	NKX3–1	Tumour suppressor gene	Deleted
8q	c-myc	Transcriptional activator	Amplified
10q	PTEN	Tumour suppressor gene	Mutated
13q	Rb	Tumour suppressor gene	Deleted
16q	E-CAD	Adhesion molecule	Deleted
Xq	AR	Androgen receptor	Amplified

KLF6

Kruppel-like factors (KLFs) comprise a group of transcription factors that appear to be involved in different biological processes including carcinogenesis. Important genetic alterations of KLF6 have been reported, including deletions and loss of expression in a minority of high grade prostate cancers.

KLF6 and NKX3.1 have not been reported as frequently as the tumour suppressor genes and have been identified because of the fact that they are within areas of frequent allelic loss in prostate tumours.

Retinoblastoma

Retinoblastoma (Rb) has been reported to be an important tumour suppressor gene in many human cancers, and prostate cancer is no exception. The disruption of the normal Rb regulatory pathway is associated with the pathogenesis of many human cancers. The Rb gene plays an important role in the G1 phase of the cell cycle. The Rb protein binds tightly to the E2F family of transcription factors. When phosphorylated, the Rb protein releases the E2F proteins, causing transcriptional activation of a variety of genes involved in cell growth. Inactivation of Rb appears to be important in neoplastic transformation, because expression of wild-type Rb in Rb negative prostate cancer lines results in loss of tumorigenicity.68 The predominant mechanism of Rb inactivation involves allelic loss or mutation, but decreased transcription of Rb has also been reported.

p53

Mutations in p53 are common in human neoplasms, but there is only a low frequency of mutation of this gene in prostate cancer. However p53 has an important role in prostate cancer progression because abnormal p53 expression is associated with bone metastases and the development of androgen independent disease. Abnormal p53 expression correlates with high histological grade, high stage, and clinical disease progression.

The p53 tumour suppressor gene product restricts entry into the synthetic phase of the cell cycle and promotes apoptosis in cells that are disorganised or have damaged DNA. Loss of normal p53 function results in uncontrolled cell growth.

The analysis of p53 expression can be difficult. The mutated p53 gene product has a longer half life, thus rendering it detectable by immunohistochemistry. However, sensitive immunohistochemical techniques may detect overexpressed normal p53. Therefore, it is more reliable to detect mutations in p53 by molecular techniques. Abnormal p53 expression is correlated with reduced survival after radical prostatectomy.

Oncogenes

c-myc and bcl-2 are well known and important oncogenes not only in prostate cancer but in many human cancers.

C-myc

Several studies have demonstrated increased myc expression in prostate cancer and a significant correlation of myc overexpression with Gleason grade. The myc oncogenes are members of the basic helix-loop-helix-leucine zipper (bHLHZ) family of transcription factors. Myc proteins act as transcriptional activators or repressors through dimerisation or with other bHLHZ family members.

Bcl-2

The bcl-2 family of genes are commonly expressed in primary and metastatic prostate cancers. bcl-2 is not expressed in the normal prostate. Proteins of the bcl-2 family play a crucial role in the regulation of apoptosis and the bcl-2 gene inhibits programmed cell death. bcl-2 has also been implicated in the development of androgen independent prostate cancer because of its increased expression in androgen independent cancers in the advanced stages of disease.

Other oncogenes have only recently been recognised, but may emerge as novel targets for molecular genetic intervention, or through modified expression may allow accurate prediction of the way in which a tumour is likely to progress.

C-Kit/Tyrosine Kinase Receptor

c-Kit (tr-Kit) is a strong activator of the Src family tyrosine kinases. In a recent study it has been reported that human trKit mRNA and protein are expressed in prostatic cancer cells. This study also describes for the first time the existence of a truncated c-Kit protein in primary tumours and shows a correlation between tr-Kit expression and activation of the Src pathway in the advanced stages of the disease.

STAT5

A signal transducer and activator of transcription, 5 (STAT5) has been identified as a crucial survival factor for prostate cancer cells. Activation of STAT5 is also associated with high histological grade of prostate cancer.

Telomerase and Telomere Shortening

Telomere length has been found to be strikingly shorter in prostate cancer, including PIN, than in normal prostate. Telomeres stabilise and protect the ends of chromosomes, but shorten because of cell division and/or oxidative damage. Critically short telomeres, in the setting of abrogated DNA damage checkpoints, have been shown to cause chromosomal instability, leading to an increase in cancer incidence as a result of chromosome fusions, subsequent breakage, and rearrangement. In normal cells, successive cycles of cellular replication result in progressive loss of telomeric sequences. Normal cells sense very short telomeres as evidence of DNA damage and exit the cell cycle. Telomerase is an enzyme that maintains telomere length by adding repetitive telomeric sequences to chromosome ends. Telomerase activity is present in most prostate cancers and not in normal prostate epithelium.

Androgen Receptor

The AR plays a crucial role in prostate cancer. AR blockade can delay the progression of prostate cancer and is used to treat patients unsuitable for radical surgery or with cancer that has spread beyond the prostate. It has been studied extensively in prostate cancer because androgens are required for the development of both the normal prostate and prostate cancer. Initially, most prostate cancers are sensitive to androgen deprivation.

However, in patients with advanced disease, most tumours progress to an androgen independent state with proliferation of cells that do not require androgens for growth.

The mechanism of acquired androgen insensitivity is unknown and has been the subject of much research, because androgen insensitive prostate cancers can no longer be treated with endocrine therapy.

Mutations, amplifications, and deletions of the AR gene and structural change in the AR protein have been postulated to cause androgen insensitivity.

"In patients with advanced disease, most tumours progress to an androgen independent state with proliferation of cells that do not require androgens for growth"

In an analysis of 44 mutant ARs from prostate cancers, seven had loss of function, three maintained wild-type function, 14 demonstrated partial function, and 20 displayed a gain in function.

References

- Harris TJ, McCormick F (2010). "The molecular pathology of cancer". Nat Rev Clin Oncol. 7 (5): 251–265. doi:10.1038/nrclinonc.2010.41. PMC 2222796. PMID 20351699
- Ogino S, Beck AH, King EE, Sherman ME, Milner DA, Giovannucci E (2012). "Ogino et Al. Respond to "the 21st century epidemiologist"". American Journal of Epidemiology. 176(8): 672–4. doi:10.1093/aje/kws229. PMC 3571249. PMID 22935516
- Molecular-diagnostics, medicine-and-dentistry: sciencedirect.com, Retrieved 17 July, 2020
- Slattery ML (October 2002). "The science and art of molecular epidemiology". Journal of Epidemiology and Community Health (Comment). 56 (10): 728–9. doi:10.1136/jech.56.10.728. PMC 1732025. PMID 12239192
- Cai, H; Caswell JL; Prescott JF (March 2014). "Nonculture Molecular Techniques for Diagnosis of Bacterial Disease in Animals: A Diagnostic Laboratory Perspective". Veterinary Pathology. 51 (2): 341–350. doi:10.1177/0300985813511132. PMID 24569613
- Molecular-pathology-of-prostate-cancer-7768044: researchgate.net, Retrieved 27 March, 2020

Permissions

All chapters in this book are published with permission under the Creative Commons Attribution Share Alike License or equivalent. Every chapter published in this book has been scrutinized by our experts. Their significance has been extensively debated. The topics covered herein carry significant information for a comprehensive understanding. They may even be implemented as practical applications or may be referred to as a beginning point for further studies.

We would like to thank the editorial team for lending their expertise to make the book truly unique. They have played a crucial role in the development of this book. Without their invaluable contributions this book wouldn't have been possible. They have made vital efforts to compile up to date information on the varied aspects of this subject to make this book a valuable addition to the collection of many professionals and students.

This book was conceptualized with the vision of imparting up-to-date and integrated information in this field. To ensure the same, a matchless editorial board was set up. Every individual on the board went through rigorous rounds of assessment to prove their worth. After which they invested a large part of their time researching and compiling the most relevant data for our readers.

The editorial board has been involved in producing this book since its inception. They have spent rigorous hours researching and exploring the diverse topics which have resulted in the successful publishing of this book. They have passed on their knowledge of decades through this book. To expedite this challenging task, the publisher supported the team at every step. A small team of assistant editors was also appointed to further simplify the editing procedure and attain best results for the readers.

Apart from the editorial board, the designing team has also invested a significant amount of their time in understanding the subject and creating the most relevant covers. They scrutinized every image to scout for the most suitable representation of the subject and create an appropriate cover for the book.

The publishing team has been an ardent support to the editorial, designing and production team. Their endless efforts to recruit the best for this project, has resulted in the accomplishment of this book. They are a veteran in the field of academics and their pool of knowledge is as vast as their experience in printing. Their expertise and guidance has proved useful at every step. Their uncompromising quality standards have made this book an exceptional effort. Their encouragement from time to time has been an inspiration for everyone.

The publisher and the editorial board hope that this book will prove to be a valuable piece of knowledge for students, practitioners and scholars across the globe.

Index

O

P

R

S

T

www.ingramcontent.com/pod-product-compliance
Lightning Source LLC
LaVergne TN
LVHW080618210926
848485LV00003B/179